T0213976

Lecture Notes in Computer Science   12685

Founding Editors

Gerhard Goos
  Karlsruhe Institute of Technology, Karlsruhe, Germany
Juris Hartmanis
  Cornell University, Ithaca, NY, USA

More information about this subseries at http://www.springer.com/series/7408

Fabiano Dalpiaz · Paola Spoletini (Eds.)

Requirements Engineering: Foundation for Software Quality

27th International Working Conference, REFSQ 2021
Essen, Germany, April 12–15, 2021
Proceedings

 Springer

Editors
Fabiano Dalpiaz 🔟
Utrecht University
Utrecht, The Netherlands

Paola Spoletini 🔟
Kennesaw State University
Kennesaw, GA, USA

ISSN 0302-9743 ISSN 1611-3349 (electronic)
Lecture Notes in Computer Science
ISBN 978-3-030-73127-4 ISBN 978-3-030-73128-1 (eBook)
https://doi.org/10.1007/978-3-030-73128-1

LNCS Sublibrary: SL2 – Programming and Software Engineering

This Springer imprint is published by the registered company Springer Nature Switzerland AG
The registered company address is: Gewerbestrasse 11, 6330 Cham, Switzerland

Preface

It is our great pleasure to welcome you to the proceedings of the 27th International Working Conference on Requirements Engineering: Foundation for Software Quality (REFSQ 2021). The REFSQ working conference series is a leading international forum for discussing requirements engineering (RE) and its role for software quality. REFSQ is an inclusive forum in which experienced researchers, PhD candidates, practitioners, and students can inform each other, learn about, discuss, and advance the state-of-the-art research and practice in RE. The first REFSQ meeting took place in 1994. The conference has been organized as a stand-alone conference since 2010 and is now well established as a premier conference series on RE, located in Europe. REFSQ 2021 was expected to be held in Essen, Germany, but the COVID-19 pandemic led the organizers to the decision to opt for a virtual conference that took place during April 12–15, 2021.

The theme of REFSQ 2021 was "Ethics as a cornerstone of Requirements Engineering", which we chose to emphasize the importance of human values, such as privacy and fairness, when designing software-intensive systems as well as the challenges that intelligent and autonomous systems pose due to the tight interplay with humans. To strengthen this perspective throughout the conference, besides some of the papers included in these proceedings, one of the keynote addresses and a few of the industry track presentations also focused on ethics.

We are pleased to present this volume comprising the REFSQ 2021 proceedings. It features 15 papers included in the technical program of REFSQ 2021, presented during the conference. These papers were selected by an international Program Committee of leading experts in RE from both academia and industry. The committee evaluated the papers via a rigorous peer-review process. This year, we received 54 abstracts, 52 of which were followed by a paper submission. We desk rejected 2 papers, leading to 50 papers sent out to the reviewers. Each paper was reviewed by three members of the REFSQ 2021 Program Committee, followed by an online discussion in EasyChair moderated by an additional member of the Program Committee who did not act as a reviewer. While most decisions were taken through the online discussion in EasyChair, we organized synchronous online meetings for 11 papers on January 14-15.

The accceptance rate of REFSQ 2021 was 28.8% (15/52). If we further analyze the details by paper category, we obtain the following:

- Technical Design (3 accepted out of 16 submissions)
- Scientific Evaluation (8 accepted out of 20)
- Research Preview (3 accepted out of 13)
- Vision (1 accepted out of 3)

The REFSQ 2021 conference was organized as a three-day symposium preceded by a day of co-located events. On Monday, the pre-conference program included three workshops (CreaRE, NLP4RE, and RE4AI), the doctoral symposium, and the graduate students' event, a special event dedicated to welcome students to the RE community prior to the beginning of a PhD. The 15 accepted papers were presented on Tuesday and Thursday. Furthermore, these two days included three keynote addresses: Katie Shilton (University of Maryland) gave a keynote on values and the role of RE as a catalyst for technology justice; Vidya Setlur (Tableau Research) focused on the embedding of AI within intelligent visual analytics tools; and Martin Glinz (University of Zurich) offered a speech on teaching requirements engineering. This third keynote paved the way for the OpenRE track, which focused on the sharing of tutorials and educational resources regarding RE. Moreover, the posters and tools track served as a way to promote early, ongoing research. Finally, the industry track program involved nine invited talks that were given on Wednesday. The papers included in all the satellite events and tracks can be found in a separate proceedings volume published via CEUR.

REFSQ 2021 would not have been possible without the engagement and support of many individuals who contributed in many different ways. As editors of this volume, we would like to thank the REFSQ Steering Committee members, in particular the chair, Kurt Schneider, and the vice-chair, Anna Perini, for their availability and excellent guidance. Special thanks go to Klaus Pohl for his long-term engagement for REFSQ. We are grateful to all the members of the Program Committee for their timely and thorough reviews of the submissions and for their time dedicated to the online discussions. We especially thank those Program Committee members who volunteered to serve in the role of gatekeeper for conditionally accepted papers. We are grateful to the chairs, who organized the various events included in REFSQ 2021, and the social media and publicity chairs Blagovesta Kostova and Oliver Karras. Finally, we would like to thank Vanessa Stricker and the team at the University of Duisburg-Essen who made the organization of the virtual conference possible and who maintained the website.

This volume consists of presentations of research results or new ideas that we hope the reader will find interesting to follow in the pursuit of his/her own work in requirements engineering.

February 2021
Fabiano Dalpiaz
Paola Spoletini

Organization

Program Committee Chairs

Fabiano Dalpiaz	Utrecht University, Netherlands
Paola Spoletini	Kennesaw State University, USA

Local Organization Chair

Vanessa Stricker	University of Duisburg-Essen, Germany

Background Organization Chair

Klaus Pohl	University of Duisburg-Essen, Germany

Industry Chairs

Anne Hess	Fraunhofer IESE, Germany
Stan Bühne	IREB, Germany

OpenRE Chairs

Neil Ernst	University of Victoria, Canada
Alessio Ferrari	CNR-ISTI, Italy

Posters and Tools Chairs

Elda Paja	IT University of Copenhagen, Denmark
Norbert Seyff	Fachhochschule Nordwestschweiz, Switzerland

Doctoral Symposium Chairs

Travis Breaux	Carnegie Mellon University, USA
Xavier Franch	Universitat Politècnica de Catalunya, Spain

Workshops Chairs

Sepideh Ghanavati	University of Maine, USA
Andreas Vogelsang	University of Cologne, Germany

Graduate Students Event Chair

Marcela Ruiz	ZHAW School of Engineering, Switzerland

Social Media and Publicity Chairs

Oliver Karras   Leibniz University Hannover, Germany
Blagovesta Kostova   EPFL, Switzerland

Proceedings Chairs

Fatma Başak Aydemir   Boğaziçi University, Turkey
Catarina Gralha   Universidade NOVA de Lisboa, Portugal

Program Committee

Raian Ali   Hamad Bin Khalifa University, Qatar
Carina Alves   Universidade Federal de Pernambuco, Brazil
Fatma Başak Aydemir   Boğaziçi University, Turkey
Muneera Bano   Deakin University, Australia
Nelly Bencomo   Aston University, UK
Richard Berntsson Svensson   Chalmers | University of Gothenburg, Sweden
Dan Berry   University of Waterloo, Canada
Tanmay Bhowmik   Mississippi State University, USA
Travis Breaux   Carnegie Mellon University, USA
Sjaak Brinkkemper   Utrecht University, Netherlands
Nelly Condori-Fernández   Universidade da Coruña, Spain
Luiz Marcio Cysneiros   York University, Canada
Maya Daneva   University of Twente, Netherlands
Joerg Doerr   Fraunhofer IESE, Germany
Alessio Ferrari   CNR-ISTI, Italy
Xavier Franch   Universitat Politècnica de Catalunya, Spain
Samuel A. Fricker   Fachhochschule Nordwestschweiz, Switzerland
Matthias Galster   University of Canterbury, New Zealand
Vincenzo Gervasi   University of Pisa, Italy
Sepideh Ghanavati   University of Maine, USA
Martin Glinz   University of Zurich, Switzerland
Michael Goedicke   University of Duisburg-Essen, Germany
Catarina Gralha   Universidade NOVA de Lisboa, Portugal
Alicia Grubb   Smith College, USA
Paul Grünbacher   Johannes Kepler University Linz, Austria
Renata Guizzardi   Universidade Federal do Espirito Santo, Brazil
Emitzá Guzmán   Vrije Universiteit Amsterdam, Netherlands
Irit Hadar   University of Haifa, Israel
Andrea Herrmann   Herrmann & Ehrlich, Germany
Jennifer Horkoff   Chalmers | University of Gothenburg, Sweden
Fuyuki Ishikawa   National Institute of Informatics, Japan
Zhi Jin   Peking University, China
Erik Kamsties   Dortmund University of Applied Sciences and Arts,
                             Germany

Alessia Knauss	Zenseact, Sweden
Eric Knauss	Chalmers | University of Gothenburg, Sweden
Kim Lauenroth	adesso AG, Germany
Seok-Won Lee	Ajou University, South Korea
Emmanuel Letier	University College London, UK
Grischa Liebel	Reykjavik University, Iceland
Nazim Madhavji	University of Western Ontario, Canada
Gunter Mussbacher	McGill University, Canada
John Mylopoulos	University of Ottawa, Canada
Nan Niu	University of Cincinnati, USA
Nicole Novielli	University of Bari, Italy
Andreas Opdahl	University of Bergen, Norway
Barbara Paech	Universität Heidelberg, Germany
Elda Paja	IT University of Copenhagen, Denmark
Oscar Pastor Lopez	Universitat Politècnica de València, Spain
Anna Perini	Fondazione Bruno Kessler, Italy
Klaus Pohl	University of Duisburg-Essen, Germany
Bjorn Regnell	Lund University, Sweden
Mehrdad Sabetzadeh	University of Ottawa, Canada
Klaus Schmid	University of Hildesheim, Germany
Kurt Schneider	Leibniz Universität Hannover, Germany
Norbert Seyff	Fachhochschule Nordwestschweiz, Switzerland
Vitor E. Silva Souza	Federal University of Espírito Santo, Brazil
Angelo Susi	Fondazione Bruno Kessler, Italy
Michael Unterkalmsteiner	Blekinge Institute of Technology, Sweden
Michael Vierhauser	Johannes Kepler University Linz, Austria
Andreas Vogelsang	University of Cologne, Germany
Yves Wautelet	Katholieke Universiteit Leuven, Belgium
Didar Zowghi	University of Technology Sydney, Australia

Additional Reviewers

Anders, Michael	Kleebaum, Anja
Dell'Anna, Davide	Koch, Matthias
Dey, Sangeeta	Rohmann, Astrid
Gupta, Sanonda Datta	Shojaifar, Alireza
Hess, Anne	Villela, Karina
Jain, Vijayanta	Waga, Masaki
Kaplan, Stephen	

Organizers

The Ruhr Institute for Software Technology

Utrecht University

KENNESAW STATE
U N I V E R S I T Y

Sponsors

International ®
Requirements
Engineering
Board

Keynotes

Practicing (Whose?) Values: Requirements Engineering as a Catalyst for Technology Justice

Katie Shilton

University of Maryland College Park, College Park, USA
kshilton@umd.edu

Abstract. Requirements engineering (RE) is a critical site for technology ethics. Technology ethics researchers have long advocated that ethical concerns be central to early design processes [1–3], and requirements engineering provides methods and processes for such reflection. RE has already begun exploring ways that social values like privacy, accessibility, and fairness can become technical requirements [4–6]. Methods have been proposed to support and systematize ethical design [7–9]. But these methods face steep challenges on the road to widespread adoption. Agile methodologies and technical cultures where ethical deliberation takes a backseat to production make values-oriented design difficult. And even after decades of debate, practitioners of values-oriented design struggle with the fundamental problem of whose values should be built into technologies which are meant to be flexible, interoperable, and global.

This talk will suggest that reframing values and ethics as explicitly about justice – considerations of power and historical oppressions – helps solve one challenge (whose values) while making the other (adoption in software engineering communities) potentially more difficult. I will then present two contrasting case studies as tools to think with and potential ways forward. The first is from my qualitative work studying independent mobile application developers. My findings suggest that there are practices already embedded in even the most informal software work which can ease the problem of adoption of ethics and justice-centered requirements engineering. The second is the history of a radically different profession: anthropology. Anthropology is a discipline that has openly grappled with power and its place in the world, and I suggest its history provides lessons for how software engineering might make reflections on power more central to development practice.

Keywords: Ethics · Justice · Software engineering.

References

1. Brey, P.A.E.: Anticipatory ethics for emerging technologies. Nanoethics. **6**, 1–13 (2012).
2. Friedman, B., Nissenbaum, H.: Bias in computer systems. In: Friedman, B. (ed.) Human Values and the Design of Computer Technology, pp. 21–40. Cambridge University Press, Cambridge and New York (1997)
3. Shilton, K.: Values and ethics in human-computer interaction. Found. Trends Human-Comput. Interact. **12**, 107–171 (2018). https://doi.org/10.1561/1100000073
4. Becker, C., Betz, S., Chitchyan, R., Duboc, L., Easterbrook, S.M., Penzenstadler, B., Seyff, N., Venters, C.C.: Requirements: the key to sustainability. IEEE Softw. **33**, 56–65 (2016). https://doi.org/10.1109/MS.2015.158
5. Anthonysamy, P., Rashid, A., Chitchyan, R.: Privacy requirements: present future. In: 2017 IEEE/ACM 39th International Conference on Software Engineering: Software Engineering in Society Track (ICSE-SEIS), pp. 13–22 (2017). https://doi.org/10.1109/ICSE-SEIS.2017.3
6. Hosseini, M., Shahri, A., Phalp, K., Ali, R.: Four reference models for transparency requirements in information systems. Requirements Eng. **23**, 251–275 (2018). https://doi.org/10.1007/s00766-017-0265-y
7. Aydemir, F.B., Dalpiaz, F.: A Roadmap for ethics-aware software engineering. In: 2018 IEEE/ACM International Workshop on Software Fairness (FairWare), pp. 15–21 (2018). https://doi.org/10.23919/FAIRWARE.2018.8452915
8. Spiekermann, S.: Ethical IT Innovation: A Value-Based System Design Approach. Auerbach Publications, Boca Raton (2015)
9. Perera, H., Mussbacher, G., Hussain, W., Shams, R.A., Nurwidyantoro, A., Whittle, J.: Continual human value analysis in software development: a goal model based approach. In: 2020 IEEE 28th International Requirements Engineering Conference (RE), pp. 192–203 (2020). https://doi.org/10.1109/RE48521.2020.00030

The Challenge(s) of Teaching Requirements Engineering

Martin Glinz

University of Zurich, Zurich, Switzerland
glinz@ifi.uzh.ch

Abstract. This extended abstract summarizes my keynote given at REFSQ 2021.

Keywords: Requirements engineering · Teaching · Education · Training

Motivation

When a new discipline such as Requirements Engineering (RE) emerges and starts spreading, we are confronted with the challenge of teaching it, both by educating students and training professionals in practice.

A Historic Perspective

The famous papers by Royce (1970) and Boehm (1976) were among the first that stated the need for treating and documenting requirements systematically. Boehm provided perspectives on systematic RE, thus giving first hints about what to teach in RE. With the advent of the special issue of IEEE TSE on RE in January 1977, the trend became clear: teach how to model requirements and the corresponding modeling languages.

Others, inspired by abstract data types and program verification, advocated teaching how to specify requirements formally, be it with logic or algebraic specifications.

The release of the IEEE standard 830 in 1984 added further teaching objectives: what are the qualities of good requirements and how to structure a requirements specification systematically.

The 1990ies and 2000s added a lot to the RE teaching agenda, for example: stakeholders, systematic requirements elicitation and validation, the importance of system context, value orientation, writing natural language requirements with phrase templates, prototyping, creativity and innovation in RE, goal-oriented RE, non-functional requirements, RE@runtime, use cases, and UML. The advent of requirements management tools added requirements management, in particular, traceability and tools.

The publication of the IREB CPRE foundation level syllabus in 2007 marked a milestone, providing a guideline for what to teach in RE on an elementary level.

More recently, the RE teaching agenda was augmented by further topics, for example, requirements evolution, collaboration, shared understanding, collecting and analyzing user feedback, and automated analysis of natural language requirements.

The growing popularity of agile development challenged RE as such and, in particular, challenged the classic, document-centric and method-focused notion of RE. As a consequence, the focus shifted toward a "principles and practices"-oriented teaching of RE, which eventually led to the new IREB CPRE syllabus version 3, published in 2020.

The Challenge(s)

Teaching modern RE is a challenge per se and it entails several challenges that teachers or trainers in RE must deal with. Below I sketch some of them that I deem important.

Motivation. Most students and many practitioners do not have practical experience in RE when attending a class or training in RE. This inevitably leads to a situation where the teacher answers questions that the students never have asked – this impedes the students' motivation to learn what they are taught.

Size. Practical exercises are a crucial ingredient of effective teaching in RE. However, such exercises suffer from a size problem: Exercises of realistic size that let students feel the pain of missing or bad RE typically are beyond of what can be done in an RE course. Conversely, exercises that fit the schedule of a course are small enough that students could solve them successfully without applying the practices that we teach.

Context. The context in which RE is applied in practice ranges from agile development and evolution of small apps to very large, plan-driven, safety-critical systems and from small, well-understood problems to complex, not or badly understood ones. It is impossible to cover such a broad field with a single, uniform set of RE methods, processes and tools – so what kinds of application contexts should we consider in teaching?

Procedures and Tools vs. Principles and Practices. Students, particularly in industrial training courses, prefer hands-on content (such as procedures and tools) that is directly applicable in practice. However, such knowledge ages rapidly and is not applicable in a broad context. Teachers, on the other hand, pursue the goal of teaching durable and broadly applicable content. This calls for teaching principles and practices.

Process. There is no universal RE process that fits most practical RE problems.

Stakeholders. Any realistic exercise in RE teaching needs stakeholders or people who act as stakeholders. However, how and from where can we get the people we need?

Consequences for Today's RE Education and Training

We cannot teach everything which is important in RE in every relevant application context. So we need to teach a set of practices, how and where to apply them, and the principles behind them. Furthermore, we should teach students how to acquire RE knowledge themselves. Case studies are a way to address the motivation and size challenges. The context and process challenges can be addressed to some extent by demonstrating how to tailor and apply RE practices in selected application contexts, using case study exercises, for example. Such demonstrations also address the students' preference for hands-on content and help mitigate the motivation challenge. Real stakeholders are possible when a course includes or is followed by a real project. Other options are simulating stakeholders with computer-based games, role plays, or scripted, multi-step case studies that describe the stakeholders' statements, behavior and reactions.

Conclusion

This is a personal and very condensed view of the topic. I do not claim any form of completeness, neither concerning the challenges nor the consequences. I hope that my keynote clarifies the historic context of RE teaching and provokes thoughts and discussions about the challenge(s) of teaching RE and how to deal with them.

What Makes Intelligent Visual Analytics Tools Really Intelligent?

Vidya Setlur

Tableau Research

Visual analysis helps people see and understand data. Effective visualizations depend on the task at hand and need to be simple, yet meaningful. While data-driven inquiry has become the norm of business practices and decision making, there is a huge untapped market of "data enthusiasts" who aren't database or computer experts; yet, they are great analytical thinkers and need tools to support their questions. There have been recent advances looking at how AI technologies can assist the analytical workflow ranging from smarter data transformations, automatic visual encodings, to supporting analytical conversation using natural language. Machine learning approaches have shown to be promising for approximating the cues for continuous learning in these systems. With a better understanding as to how users explore data in their flow of analysis, a natural question is can user behavior be applied as a set of engineering requirements to developing smarter tools? That is, can people doing analysis be supported or even replaced by more intelligent tools? In this talk, I will explore this question.

Contents

From Software to Systems and Services

Analysts' Competence and Training

Natural Language Processing and Machine Learning

Is Requirements Similarity a Good Proxy for Software Similarity? An Empirical Investigation in Industry

Muhammad Abbas[1,2]($\boxtimes$), Alessio Ferrari[3], Anas Shatnawi[4],
Eduard Paul Enoiu[2], and Mehrdad Saadatmand[1]

[1] RISE Research Institutes of Sweden, Västerås, Sweden
{muhammad.abbas,mehrdad.saadatmand}@ri.se
[2] Mälardalens University, Västerås, Sweden
{muhammad.abbas,eduard.enoiu}@mdh.se
[3] CNR-ISTI, Pisa, Italy
alessio.ferrari@isti.cnr.it
[4] Berget-Levrault, Montpellier, France
anas.shatnawi@berget-levrault.com

Abstract. [**Context and Motivation**] Content-based recommender systems for requirements are typically built on the assumption that similar requirements can be used as proxies to retrieve similar software. When a new requirement is proposed by a stakeholder, natural language processing (NLP)-based similarity metrics can be exploited to retrieve existing requirements, and in turn identify previously developed code. [**Question/problem**] Several NLP approaches for similarity computation are available, and there is little empirical evidence on the adoption of an effective technique in recommender systems specifically oriented to requirements-based code reuse. [**Principal ideas/results**] This study compares different state-of-the-art NLP approaches and correlates the similarity among requirements with the similarity of their source code. The evaluation is conducted on real-world requirements from two industrial projects in the railway domain. Results show that requirements similarity computed with the traditional *tf-idf* approach has the highest correlation with the actual software similarity in the considered context. Furthermore, results indicate a *moderate positive* correlation with Spearman's rank correlation coefficient of more than 0.5. [**Contribution**] Our work is among the first ones to explore the relationship between requirements similarity and software similarity. In addition, we also identify a suitable approach for computing requirements similarity that reflects software similarity well in an industrial context. This can be useful not only in recommender systems but also in other requirements engineering tasks in which similarity computation is relevant, such as tracing and categorization.

Keywords: Requirements similarity · Software similarity · Correlation

This work has been supported by and received funding from the ITEA3 XIVT, and KK Foundation's ARRAY project.

© Springer Nature Switzerland AG 2021
F. Dalpiaz and P. Spoletini (Eds.): REFSQ 2021, LNCS 12685, pp. 3–18, 2021.
https://doi.org/10.1007/978-3-030-73128-1_1

1 Introduction

Recommender systems have been widely studied in requirements engineering (RE) [14,19,28], and several diverse applications of this paradigm have been proposed in the literature. These include stakeholder recommendation for requirements discussions [8], refactoring recommendation based on feature requests [27] and also bid management [14]. One typical application scenario of recommender systems in RE is related to *requirements retrieval* [9,20] in reactive software product line engineering (SPLE) [22,35]. With SPLE, companies manage software reuse in a structured way to satisfy multiple variations of customer requirements while minimizing development effort [29]. Specifically, in a reactive SPLE context [22], when a new requirement is proposed, the requirements analyst looks for reuse opportunities and compares the new proposal with existing requirements in order to adapt their previously developed models and implementations. This can be supported by content-based recommender systems [24], which, given a new requirement, return the most similar ones in a historical database of product releases, together with the associated artifacts. The rationale of the approach is that similar requirements can be used as proxies to retrieve similar software, i.e., code that can be adapted with little effort to address the new needs. Different NLP techniques exist to compute semantic requirements similarity, and the recent emerging of novel NLP language models provides promising options [38]. However, it is unclear to which extent requirements similarity implies software similarity and what are the most effective techniques to support requirements similarity computation in a way that is optimized for code retrieval. This paper aims to empirically study the problem in the context of the requirements of Bombardier Transportation AB (BT), a world-leading railway company. The main objective of this study is to improve the requirement-based software retrieval process in the studied setting. To study the relationship between requirements similarity and software similarity, we consider 254 real-world requirements related to two Power Propulsion Control (PPC) projects. We consider different state-of-the-art language models to semantically represent the requirements and support similarity computation, namely the traditional *tf-idf* [25], and the more advanced Doc2Vec [23], FastText [5], and Bidirectional Encoder Representations from Transformers (BERT) [10]. Surprisingly, our results show that, in our context, the traditional *tf-idf* model is the one that leads to the highest correlation with the software similarity, computed with JPLag [30]. Furthermore, we show that, with the exception of the Doc2Vec case, the correlation between requirements similarity and code similarity is moderate. This provides some evidence that similar implementations realize similar requirements in the context of the considered case study but also suggests that there is further space for research about novel methods to retrieve similar software that goes beyond requirements similarity.

The rest of the paper is organized as follows. Related work is discussed in Sect. 2. Section 3 discusses the research design, with context, research questions, and procedures. In Sect. 4, we present the results, and in Sect. 5 we discuss the main takeaway messages. Threats to validity are presented in Sect. 6. We conclude the paper and draw future directions in Sect. 7.

2   Related Work

In software engineering, several approaches rely on similarity measurements to analyze relationships between different software artifacts. Typical goals include feature identification [39], feature location [12], architecture recovery [35], reusable service identification [34] and clone detection [36]. In the RE field, similarity computation normally involves using NLP techniques to represent the requirements [38], as these are typically written in NL [15]. Computation of similarity is key for many typical requirements management tasks, including traceability [6,17,18], identification of equivalent requirements [13], clustering [3], and also recommender systems based on Information Retrieval (IR) approaches [2,8,9,11,14,19,27,28,32]. As our research is focused on this latter group of applications, we will compare our work with representative ones in this area. One of the seminal contributions is the work by Natt och Dag *et al.* [9], where the *tf-idf* language model and cosine similarity are used to support retrieval of previous requirements on a large industrial dataset. Dumitru *et al.* [11] propose an approach for feature recommendation based on online product descriptions, with the support of association rule mining and *k*NN clustering. This type of clustering is also used by Castro-Herrera *et al.* [8], who proposes to recommend relevant stakeholders to requirements discussion forums based on their expertise. The OpenReq EU project [14,28] aims to take a more holistic perspective, with recommendations in elicitation, specification, and analysis, and also includes a proposal for bid management. Similarly to our work, the researchers plan to use content-based recommender systems for requirements and adopt vector-space language models to support similarity computation. On a different note, but still using *tf-idf* to support similarity computation, Nyamawe *et al.* [27] recommend refactoring based on new feature requests. Finally, in a recent contribution [2], we used requirements descriptions to recommend the reuse of their implementation for new requirements.

Compared to our previous work [2], which was dedicated to the whole task of software reuse, the current investigation is explicitly focusing on exploring the relationship between requirements similarity and the actual software similarity. With respect to other related studies, the work presented in this paper is the first one that, while focusing on requirements and software similarity, compares the most recent state-of-the-art NLP techniques to support similarity computation and applies these techniques in an industrial context. This is particularly relevant also for the whole NLP for the RE field, as the recent survey of Zhao *et al.* [38] clearly highlights limited experimentation with advanced NLP techniques in RE research.

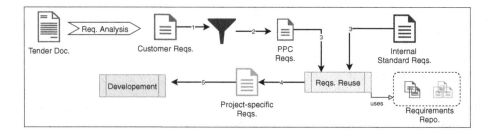

Fig. 1. The overall process of receiving requirements from a customer

3   Study Design

This section outlines the research method used to obtain the results. This work can be regarded as an *exploratory* case study [33], oriented to understand the relationship between requirements and their associated code and exploit this relationship for software retrieval in the specific context of a railway company. We designed this study following the guidelines of Runeson *et al.* [33] for conducting and reporting case studies by providing an overview of the context, our objective, and research questions, followed by the data collection and analysis procedures.

3.1   Study Context

We have studied the PPC software development team of BT. In this team, the software is developed by reusing existing components from the assets base [1]. The development of a new product starts after receiving customer require-ments from different teams at the company. Since the system is a safety-critical software-intensive system, the requirements for all existing products can be traced to the source code. The team consists of more than 140 employees, devel-oping safety-critical products, and thus the requirements have to be dealt with in detail. Therefore, all the team members participate in the requirements engineer-ing activities. As shown in Fig. 1, requirement analysis and elicitation activities are performed on tender documents to extract the customer requirements. The customer requirements relevant to the propulsion system are received by this team. The input requirements (PPC reqs.) are internalized by reusing standard internal domain requirements and existing requirements from other projects. This results in project-specific internal requirements to be implemented.

To support reuse, the engineers also conduct reuse analysis to identify exist-ing similar customer requirements and, by exploiting traceability information, identify existing software components that could be reused to realize the new requirements. However, this process is heavily dependent on the experience of engineers and is time-consuming. Currently, the process is being automated with a recommender system called VARA [2]. Like most RE recommender systems, VARA is also based on the assumption that similar requirements can be used as proxies to retrieve similar software.

3.2   Objective and Research Questions

Our main goal is to improve the software retrieval process in the studied setting. To this end, we need first to check the typical assumption of content-based recommender systems for requirements, i.e., that similar requirements can be used as proxies to retrieve similar software. In other terms, we want to check if a relationship can be identified between requirements similarity and software similarity so that similar requirements can be assumed to point to similar code. Then, we want to check which NLP approach performs best in exploiting this similarity. To achieve these objectives, we define the following research question.

> *RQ: To which extent can we use requirements similarity, automatically computed through different language models, as a proxy for software similarity?*

This research question aims at exploring the relationship between requirements similarity and software similarity. Language models are commonly used to compute the similarity among requirements. Therefore, this research question also aims to identify the most effective language model in our specific case for computing requirements similarity that correlates well with the software similarity and can be better exploited in the given setting.

The case under study is the relationship between requirement similarity and software similarity in the considered industrial setup. The unit under analysis in our case are the projects developed in the Power Propulsion Control software at Bombardier Transportation.

3.3   Data Collection

We collected data from two projects at Bombardier Transportation AB, developed by the Power Propulsion Control software team. Due to limited access to the company's repository, the projects were selected based on convenience by a company's project manager. The requirement documents were subjected to cleaning to remove all non-requirements such as headings and definitions. This resulted in a final set of 254, selected out of 265 entries. In data collection, one project manager from the company was involved in validating our procedure. Table 1 outlines the data about two projects with information on requirements and lines of code.

Table 1. Summary of the selected requirements with and without stop-words

Project	Reqs.	With		Without		SLOC
-	-	**Words**	**AVG. Words**	**Words**	**AVG. Words**	-
A	112	5823	51.9	3308	29.5	53.7K
B	142	10736	75.6	6478	45.6	61K
Total	254	16559	63.7	9786	37.5	114.7K

We conducted the investigation for our dataset, both with and without stop-words. This is because some language models can utilize stop-words, suffix, and prefix information for learning. We use a pre-processing pipeline to remove all the English stop-words and lemmatize the words of the requirements to their roots. An example requirement from the PURE dataset before and after pre-processing is shown as follows [16].

Before Pre-Processing: The number of block movement in incremented of 1. The difference of time of the block movement and the previous recorded time is recorded.

After Pre-Processing: number block movement incremente 1 difference time block movement previous record time record

---

In the studied projects, the requirements are realized in Simulink models, and code is generated from the models for deployment. Besides, there are not many available tools for computing similarity between Simulink models. Therefore, to mimic the studied setting, we used Simulink Embedded Coder[1] with MinGW64 gmake tool-chain to generate code from the models. The related code realizing each requirement was traced and moved to directories tagged with the requirement's identifiers.

3.4 Language Models for Requirements Similarity

Language models are used to derive feature vectors from the requirements' text. Various similarity metrics are used on the vectors to compute similarity among them. The cosine similarity metric is based on the cosine angle between the vectors and is heavily used in the area of NLP. The effectiveness of the similarity computed with cosine is heavily dependent on the choice of language model used for computing feature vectors. In addition, some language models are sensitive to pre-processing, such as removal of stop-words and lemmatization. This is why we selected some of the most seminal language models and fed them the dataset with and without pre-processing applied. Particularly, we considered tf-idf (TF), Doc2Vec (DW), FastText (FT), and BERT. In addition, to see the effect of pre-processing, we combined these language models with pre-processing (pTF, pDW, pFT, and pBERT). Note that for DW, FT, and BERT, the hyper-parameters are not in our control and are coming from the original pre-trained models. In our case, the input to each language model is the requirements from two projects, and the output is vectors of requirements. Given the total number of requirements, we select the top 50 similar pairs of requirements using cosine similarity to fulfill the sample size requirement. A pair is created by retrieving the most similar requirement from project B for each requirement in project A. The similarity between each pair of requirements' vectors is calculated using the cosine similarity metric implementation available in scipy [31]. In this sub-section, we first present the pre-process pipeline, then the different language models used to generate vectors to compute the similarity between requirement pairs.

---

[1] The option "optimize for traceability" was selected in Embedded Coder.

Pre-Process. The pre-process pipeline takes the requirements text and removes English stop-words from it. After the removal of the stop-words, each token of the requirements text is tagged with Part-of-speech (POS) tags to guide the lemmatization. The pre-trained spaCy model[2] is used to lemmatize the text of the requirement. The output of this pipeline is the pre-processed text of the requirement. The dataset before and after pre-processing is shown in Table 1. In the remainder of this section, the names of language models starting with "p" are the model variants where pre-processing is applied.

TF is based on the *tf-idf* score from IR. TF extracts term-matrix from the input requirements where the terms are treated as features, and the frequencies are treated as values of the features. Minimum and maximum term frequencies can be defined to drop irrelevant features such as potential stop-words. The matrix also considers the co-occurring terms (n-grams) as features. The term matrix is usually of very high dimensions, and thus dimensionality reduction techniques are used to select the top features from the matrix. Such an approach is useful in cases where the requirements share common terms. In our case, the model is configured to build the term-document matrix on project B and then uses Principal Component Analysis (PCA) [21] to select the top features based on the explained variance of 95%from the matrix. The minimum and maximum document frequencies are set to 6 and 0.5, respectively. We consider n-grams ranging from 1 to 8.

DW is based on the Word2Vec approach, where every word in a document is mapped to a vector of real numbers using a neural network. The vectors are concatenated to get vectors for the entire document, preserving the contextual and semantic information For example, words like "simple" and 'easy" would result in similar vectors. This helps in inferring feature vectors of fixed-length for a variable length of requirements. In our case, the pre-trained Doc2Vec model available in Gensim data[3] is used. The model has a vector size of 300, with a minimum frequency set to 2. The model is trained on the English Wikipedia documents resulting in a vocabulary size of 35,556,952.

FT is another model based on Word2Vec, where instead of learning word vectors directly, it utilizes the character level n-grams. For example, the word "run" would be divided into n-grams such as "ru," "run," "un". Such a model is useful in cases where shorter words are used. In addition, FastText also understands suffixes (such as verb ending) and prefixes (such as unhappy, where *un* is the prefix) better because it utilizes character-level information. In our case, we use the pre-trained FT model available in Gensim data. The model has a vector size of 100 with a minimum frequency set to 1. The model is trained on the English Wikipedia documents on the sub-word-level, resulting in a vocabulary size of 2,519,370. Both FT and DW are based on the skip-gram neural network architecture [26], known for contextual word prediction.

BERT is a recent breakthrough in language understanding research. It is a bi-directional model based on the Transformer encoder architecture that also

---

[2] https://spacy.io/.
[3] https://github.com/RaRe-Technologies/gensim-data.

considers positional and contextual information of words. BERT is known for the so-called contextual embedding and is trained on BooksCorpus and the English Wikipedia with 2,500M words. Such a model could be handy for capturing the semantic of the requirements. In our case, we use the uncased pre-trained BERT model by Google Research [10]. The model has 12 layers and a vector size of 768. We use the BERT implementation available in BERT-as-a-service[4].

3.5   Software Similarity Pipeline

Our software similarity pipeline takes pairs of requirement's identifiers as input. It copies each pair's code to separate folders[5]. The pipeline then uses JPLag to compute the similarity between the pair of source code. To compute the similarity between the source code of the two requirements, we use the JPLag's Java ARchive (JAR) with C/C++ as a language parameter [30]. JPLag was originally designed to detect plagiarism in students' assignments and thus is able to detect semantically similar code. Note that JPLag ignores code comments and white-spaces and scans and parses the input programs to convert the programs into string tokens. JPLag then uses a greedy version of string tiling algorithm to compute the similarity between the tokens of the source code. The similarity number is basically the percentage of similar tokens in the pairs of source code. The output of this pipeline is the software similarity values between 0 and 100, later converted to range between 0 and 1 for the input pairs.

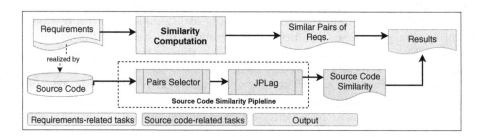

Fig. 2. Execution procedure overview

3.6   Execution

Figure 2 shows a high-level view of the execution procedure followed to obtain the results. We started with two requirement documents as input to all the language models presented in Sect. 3.4. Each language model outputs vectors of the requirements that are used to select the 50 most similar pairs of requirements

---

[4] Xiao Han, https://github.com/hanxiao/bert-as-service.
[5] In our case, each folder for a pair contains two sub-folders with code of each requirement.

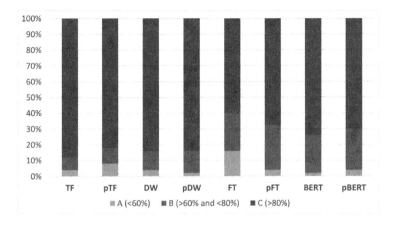

Fig. 3. Software similarity distribution in the top 50 similar requirement pairs

based on cosine similarity. For each model, the `Pairs Selector` searches, selects and structures the code of the requirements for JPLag. The pipeline then uses JPLag to compute the similarity between each input pair of the source code and produces the software similarity values for each language models' result.

3.7 Data Analysis

First, we visualize the data in bar and scatter plots to provide descriptive statistics on the software similarity percentages among the identified pairs by using each language model. Then, we apply the correlation analysis to quantify the relationship between the two variables using R Studio[6]. As our data are not normally distributed and we do not assume any linear correlation between the variables, we use Spearman's rank correlation coefficient test.

4   Results

In this section, we quantitatively answer our posed research question. First, we present the descriptive statistics, and then we present the correlation analysis.

Descriptive Statistics. To understand the results, we divided the similar pairs of the requirements—computed based on different language models—against the actual software similarity into three classes. The first class represents the cases where the retrieved software shares less similarity (<60% software similarity, A). The second class represents cases where the retrieved software share moderate similarity (between 60 and 80% between the software of the pairs, B), finally, third class represent cases where the retrieved software shares high similarity (>80% similarity between the software of the pairs, C). The above classes are

---

[6] RStudio, Available online, https://rstudio.com/.

defined to show the extent to which requirements similarity can be used to recommend requirements-based code reuse. Figure 3 shows the distribution of software similarity among the top 50 similar pairs of requirements based on each language model. As it can be seen, in all cases, in at-least 60% of the pairs, the software similarity stays above 80% (in class C).

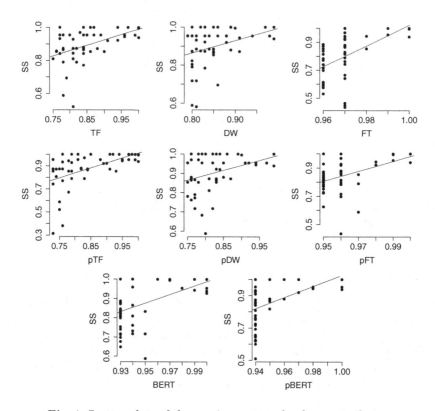

Fig. 4. Scatter plots of the requirements and software similarity

In addition, Fig. 4 presents a holistic view of the association between the requirements similarity and software similarity. The requirements similarity (on X-Axis) is calculated using different language models. The software similarity is plotted on Y-Axis and is calculated using our JPLag-based pipeline, shown in Fig. 2. The blue line is the trendline between the two variables, giving insights into the relationship between them. In all cases, as can be seen from the trendlines, there could be a positive association between the two variables. Besides, we also visualize the interquartile range (IQR), mean, and outliers in our variables in Fig. 5. As can be seen from Fig. 5, the software similarity for most requirement pairs stays above 70%.

Correlation Analysis. We applied correlation analysis to quantify the relationship and find the most suitable approach toward requirements similarity computation.

We measure the correlation between the similarity of the top 50 most similar pairs of the requirements and their source code similarity. We choose the top 50 pairs because it is a suitable number for a sample size (for applying statistical tests) and, at the same time, not a large number of pairs compared to the total requirements.

Table 2 show the results of Spearman's rank correlation. The `p-value` indicates the significance of the obtained results. The `rho` column is the correlation coefficient, which ranges from -1 to 1. As it can be observed, there is a positive association between the requirements similarity and software similarity for all the language models.

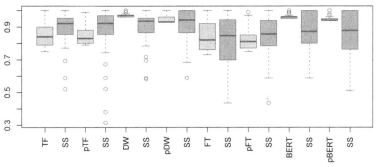

Fig. 5. Requirements Similarity (blue) and their corresponding Software Similarity (SS, purple) for all pipelines (Color figure online)

Table 2. Spearman's rank Correlation Results with Moderate correlation in bold text. The best pipeline (pTF) is also reported in bold.

	TF	pTF	DW	pDW	FT	pFT	BERT	pBERT
rho	**0.5089**	**0.5927**	0.2642	0.3104	**0.5718**	0.4676	0.3865	**0.5575**
p-value	0.0001	5.753e−06	0.0636	0.0282	1.439e−05	0.0006	0.0055	2.594e−05

5   Discussion

From the results shown in Fig. 3, it can be seen that even in worst cases, the requirements-based code retrieval would result in retrieving code with a high software similarity (that is more than 80%), which can be therefore a good candidate for reuse. Based on the descriptive statistics, we can make the following conclusion.

> *Requirement-level similarity can be used as proxy for retrieving relevant software (sharing at-least 80% software similarity) for reuse in at-least 60% of the cases.*

In addition, the trendlines in Fig. 4 also shows that the results from all the language models could have a positive association with software similarity. However, in some cases these language models can produce inaccurate results. As it can be seen in Fig. 5, there are some outliers in the retrieved software. Besides it can also be seen from Fig. 5, the variance in the software similarity across the pairs is high in case of FT and BERT, suggesting that these models tend to capture more nuanced semantic similarities in requirements, which may point to more fine-grained variations of the software. For these language models, the minimum software similarity can also be quite low, therefore indicating that the nuanced similarities in requirements can also lead to software that cannot be easily reused. These more semantically-laden representation may be more appropriate for tasks other than code retrieval, such as, e.g., requirements-to-requirements tracing, where dependencies tend to go far beyond lexical aspects. Figure 5 also shows that similarity ranges largely vary between language models (e.g., BERT and DW have very limited range with respect to the others). This suggests that having a code-retrieval system that is based on thresholds over the similarity values (e.g., consider software with requirements similarity higher than 75%) may not be the most appropriate solution.

The correlation analysis (presented in Table 2) shows that for all language models, we were able to find a positive correlation between requirements similarity and software similarity. In particular, there is a moderate positive correlation between the requirements similarity computed with tf-idf, FastText and BERT (shown in bold text in Table 2). Results also show that pre-processing improves the correlation for all language models except FastText.

> *Our results indicate that term-frequency inverse document frequency (tf-idf)-based language model with pre-processing shows a moderately positive correlation (with rho of 5.92) to software similarity.*

Surprisingly, the decades-old *tf-idf* performs better than the new state-of-the-art language models. This can be explained by the limited vocabulary and high similarity of terms used in the requirements of the two projects, as typical in the RE domain [16], where synonyms are not recommended, and company practices encourage uniform terminology. In tasks where requirements might be sharing fewer terms—e.g., in case of comparison between high-level customer requirements and low-level specifications—, the benefit of language models capturing semantics, such as BERT, could be more evident. The worst performance is obtained with Doc2Vec. This language model works well with long documents and might not be a good candidate for RE tasks, as single requirements are typically short, but maybe beneficial in contexts where the comparison is performed between entire requirements documents.

It is worth remarking that our final objective is to improve requirements-based software reuse. At this stage, our results can be used to build recommender systems that use requirements similarity to retrieve candidate software for reuse. The actual in-field assessment of the quality of the retrieval—or, in other terms, the answer to the question: *does the retrieved software satisfy my requirement?*—will need to be addressed with the involvement of human operators.

6 Threats to Validity

In this section, we present validity threats according to Runeson et al. [33].

We based the problem of software retrieval for reuse at the requirements-level and provided empirical evidence on the association between requirements similarity and software similarity. In our procedure, we used pre-trained models that are heavily dependent on the quality of the training dataset. The quality of the results might differ if different pre-trained language models are considered. To mitigate potential threats to construct validity, we selected a diverse set of approaches (see Sect. 3.4) to represent the semantics of the requirements. We did not consider similarity as assessed by human subjects, as our goal is to use language models for automatic similarity computation. However, different results may emerge if human subjects are involved in the assessment.

To mitigate potential internal validity threats, we followed standard procedure and open source implementations. In addition, we also involved researchers from diverse backgrounds to validate the study design and execution. Finally, we also involved a technical project manager at the company in validating our data collection procedure.

Our results are based on data provided by one company using a data set of two projects developed by a team. We do not claim the generalizability of our results beyond this context. In addition, our results are only limited to one level of abstraction since we do not consider multiple levels of requirement refinement. Considering guidelines for case-based generalization [37], these results might be applicable to similar contexts, where similar RE practices are followed. Further studies are needed on other abstraction levels of requirements and in different companies and domains to generalize the results.

Finally, we address the threats to the reliability of our results by providing enough details on the experimental setup and implementation. In addition, we also provide the R script and the similarity values between the pairs for replication purposes[7].

7 Conclusion and Future Work

Content-based recommender systems for code retrieval typically use requirements as queries to identify previously developed requirements, and in turn, reuse their implementation. These systems take the operational assumption that

---

[7] Replication package, https://doi.org/10.5281/zenodo.4275388.

similar requirements can be used as proxies to retrieve similar code that can be reused with limited adaptation. This paper presents an empirical investigation on the relationship between requirements similarity and code similarity in the context of a railway company. The goal of the work is to explore to which extent similar requirements can be considered as a proxy to retrieve similar code. We consider two related projects in the company. We use different NLP-based language models to represent the requirements and support similarity computation. Given similar requirements, we identify the associated code, and we compute code similarity with JPLag. Our analysis shows that the correlation between requirements and code similarity is moderately positive, even in the best case. So, a relationship exists between the two, but there is also a need for further research on language models and similarity measurement approaches that can better reflect software similarity. In our specific case, the language model that reflects software similarity better is the traditional *tf-idf*.

Future work will consider a broader set of possible application scenarios of recommender systems for code reuse. Avenues that we plan to explore include: (1) considering the original tender requirements, and identify the relationship with existing requirements and associated software, to support early evaluation during bid proposal (2) considering feature or refactoring requests as input queries, to support change impact analysis [4,7] (3) consider other companies and domains other than railways to increase external validity of the results (4) involve domain experts in the assessment of similarity measurements, as well as in the empirical evaluation of requirements-based software retrieval for reuse (5) identify when a specific language model is more appropriate to compute similarity, given the types of relationship between the format of the queries accepted by the recommender system, the characteristics of the requirements (e.g., high- *vs* low-level, functional *vs* quality), and the type of activity that is expected to be performed with the retrieved software, which can be reused, but also correct, remove, end even validate. Indeed, similarity measures and code retrieval can also be exploited to identify incorrectly traced software or missing trace links [17,18], as well as potentially tacit requirements that are implemented in the software but are not specified.

References

1. Abbas, M., Jongeling, R., Lindskog, C., Enoiu, E.P., Saadatmand, M., Sundmark, D.: Product line adoption in industry: an experience report from the railway domain. In: Proceedings of the 24th ACM Conference on Systems and Software Product Line: Volume A - Volume A. SPLC 2020. ACM, New York (2020)
2. Abbas, M., Saadatmand, M., Enoiu, E., Sundamark, D., Lindskog, C.: Automated reuse recommendation of product line assets based on natural language requirements. In: Ben Sassi, S., Ducasse, S., Mili, H. (eds.) Reuse in Emerging Software Engineering Practices, pp. 173–189. Springer, Cham (2020). https://doi.org/10.1007/978-3-030-64694-3_11
3. Arora, C., Sabetzadeh, M., Briand, L., Zimmer, F.: Automated extraction and clustering of requirements glossary terms. Trans. Soft. Eng. **43**(10), 918–945 (2016)

4. Arora, C., Sabetzadeh, M., Goknil, A., Briand, L.C., Zimmer, F.: Change impact analysis for natural language requirements: an NLP approach. In: International Requirements Engineering Conference (RE), pp. 6–15. IEEE (2015)
5. Bojanowski, P., Grave, E., Joulin, A., Mikolov, T.: Enriching word vectors with subword information. Trans. Assoc. Comput. Linguist. **5**, 135–146 (2017)
6. Borg, M., Runeson, P., Ardö, A.: Recovering from a decade: a systematic mapping of information retrieval approaches to software traceability. Empir. Softw. Eng. **19**(6), 1565–1616 (2014). https://doi.org/10.1007/s10664-013-9255-y
7. Borg, M., Wnuk, K., Regnell, B., Runeson, P.: Supporting change impact analysis using a recommendation system: an industrial case study in a safety-critical context. IEEE Trans. Soft. Eng. **43**(7), 675–700 (2016)
8. Castro-Herrera, C., Cleland-Huang, J., Mobasher, B.: Enhancing stakeholder profiles to improve recommendations in online requirements elicitation. In: International Requirements Engineering Conference, pp. 37–46. IEEE (2009)
9. Natt och Dag, J., Regnell, B., Gervasi, V., Brinkkemper, S.: A linguistic-engineering approach to large-scale requirements management. IEEE Softw. **22**(1), 32–39 (2005)
10. Devlin, J., Chang, M.W., Lee, K., Toutanova, K.: Bert: Pre-training of deep bidirectional transformers for language understanding. arXiv preprint arXiv:1810.04805 (2018)
11. Dumitru, H., et al.: On-demand feature recommendations derived from mining public product descriptions. In: International Conference on Software Engineering, pp. 181–190 (2011)
12. Eyal-Salman, H., Seriai, A.D., Dony, C.: Feature-to-code traceability in a collection of software variants: combining formal concept analysis and information retrieval. In: 2013 IEEE 14th International Conference on Information Reuse & Integration (IRI), pp. 209–216 (2013)
13. Falessi, D., Cantone, G., Canfora, G.: Empirical principles and an industrial case study in retrieving equivalent requirements via natural language processing techniques. Trans. Softw. Eng. **39**(1), 18–44 (2011)
14. Felfernig, A., Falkner, A., Atas, M., Franch, X., Palomares, C.: OpenReq: recommender systems in requirements engineering. In: RS-BDA, pp. 1–4 (2017)
15. Fernández, D.M., et al.: Naming the pain in requirements engineering. Empir. Softw. Eng. **22**(5), 2298–2338 (2017)
16. Ferrari, A., Spagnolo, G.O., Gnesi, S.: Pure: a dataset of public requirements documents. In: 2017 IEEE 25th International Requirements Engineering Conference (RE), pp. 502–505 (2017). https://doi.org/10.1109/RE.2017.29
17. Gervasi, V., Zowghi, D.: Supporting traceability through affinity mining. In: International Requirements Engineering Conference (RE), pp. 143–152. IEEE (2014)
18. Guo, J., Cheng, J., Cleland-Huang, J.: Semantically enhanced software traceability using deep learning techniques. In: International Conference on Software Engineering (ICSE), pp. 3–14. IEEE (2017)
19. Hariri, N., Castro-Herrera, C., Cleland-Huang, J., Mobasher, B.: Recommendation systems in requirements discovery. In: Robillard, M.P., Maalej, W., Walker, R.J., Zimmermann, T. (eds.) Recommendation Systems in Software Engineering, pp. 455–476. Springer, Heidelberg (2014). https://doi.org/10.1007/978-3-642-45135-5_17
20. Irshad, M., Petersen, K., Poulding, S.: A systematic literature review of software requirements reuse approaches. IST J. **93**, 223–245 (2018)

21. Jolliffe, I.T., Cadima, J.: Principal component analysis: a review and recent developments. Philos. Trans. Royal Soc. A: Math. Phys. Eng. Sci. **374**(2065), 20150202 (2016)

22. Krueger, C.W.: Easing the transition to software mass customization. In: van der Linden, F. (ed.) PFE 2001. LNCS, vol. 2290, pp. 282–293. Springer, Heidelberg (2002). https://doi.org/10.1007/3-540-47833-7_25

23. Le, Q., Mikolov, T.: Distributed representations of sentences and documents. In: International Conference on Machine Learning, pp. 1188–1196 (2014)

24. Lops, P., de Gemmis, M., Semeraro, G.: Content-based recommender systems: state of the art and trends. In: Ricci, F., Rokach, L., Shapira, B., Kantor, P.B. (eds.) Recommender Systems Handbook, pp. 73–105. Springer, Boston, MA (2011). https://doi.org/10.1007/978-0-387-85820-3_3

25. Manning, C.D., Schütze, H., Raghavan, P.: Introduction to Information Retrieval. Cambridge University Press, Cambridge (2008)

26. Mikolov, T., Chen, K., Corrado, G., Dean, J.: Efficient estimation of word representations in vector space (2013)

27. Nyamawe, A.S., Liu, H., Niu, N., Umer, Q., Niu, Z.: Automated recommendation of software refactorings based on feature requests. In: International Requirements Engineering Conference (RE), pp. 187–198. IEEE (2019)

28. Palomares, C., Franch, X., Fucci, D.: Personal recommendations in requirements engineering: the OpenReq approach. In: Kamsties, E., Horkoff, J., Dalpiaz, F. (eds.) REFSQ 2018. LNCS, vol. 10753, pp. 297–304. Springer, Cham (2018). https://doi.org/10.1007/978-3-319-77243-1_19

29. Pohl, K., Böckle, G., van Der Linden, F.J.: Software Product Line Engineering: Foundations, Principles and Techniques. Springer, Heidelberg (2005)

30. Prechelt, L., Malpohl, G., Philippsen, M., et al.: Finding plagiarisms among a set of programs with JPlag. J. UCS **8**(11), 1016 (2002)

31. Řehůřek, R., Sojka, P.: Software framework for topic modelling with large corpora. In: Proceedings of the LREC 2010 Workshop on New Challenges for NLP Frameworks, pp. 45–50. ELRA, May 2010

32. Robillard, M.P., Maalej, W., Walker, R.J., Zimmermann, T. (eds.): Recommendation Systems in Software Engineering. Springer, Heidelberg (2014). https://doi.org/10.1007/978-3-642-45135-5

33. Runeson, P., Höst, M.: Guidelines for conducting and reporting case study research in software engineering. Empir. Softw. Eng. **14**(2), 131–164 (2009)

34. Shatnawi, A., Seriai, A., Sahraoui, H., Ziadi, T., Seriai, A.: Reside: reusable service identification from software families. JSS **170** (2020)

35. Shatnawi, A., Seriai, A.D., Sahraoui, H.: Recovering software product line architecture of a family of object-oriented product variants. J. Syst. Softw. **131**, 325–346 (2017)

36. White, M., Tufano, M., Vendome, C., Poshyvanyk, D.: Deep learning code fragments for code clone detection. In: International Conference on Automated Software Engineering (ASE), pp. 87–98. IEEE (2016)

37. Wieringa, R., Daneva, M.: Six strategies for generalizing software engineering theories. Sci. Comput. Program. **101**, 136–152 (2015)

38. Zhao, L., et al.: Natural language processing (NLP) for requirements engineering: A systematic mapping study. arXiv preprint arXiv:2004.01099 (2020)

39. Ziadi, T., Frias, L., da Silva, M.A.A., Ziane, M.: Feature identification from the source code of product variants. In: 2012 16th European Conference on Software Maintenance and Reengineering, pp. 417–422. IEEE (2012)

Automatic Detection of Causality in Requirement Artifacts: The CiRA Approach

Jannik Fischbach[1]([✉]), Julian Frattini[2], Arjen Spaans[1], Maximilian Kummeth[1], Andreas Vogelsang[3], Daniel Mendez[2,4], and Michael Unterkalmsteiner[2]

[1] Qualicen GmbH, Munich, Germany
{jannik.fischbach,arjen.spaans,maximilian.kummeth}@qualicen.de
[2] Blekinge Institute of Technology, Karlshamn, Sweden
{julian.frattini,daniel.mendez,michael.unterkalmsteiner}@bth.se
[3] University of Cologne, Cologne, Germany
vogelsang@cs.uni-koeln.de
[4] fortiss GmbH, Munich, Germany
mendez@fortiss.org

Abstract. [**Context & motivation:**] System behavior is often expressed by causal relations in requirements (e.g., *If event 1, then event 2*). Automatically extracting this embedded causal knowledge supports not only reasoning about requirements dependencies, but also various automated engineering tasks such as seamless derivation of test cases. However, causality extraction from natural language (NL) is still an open research challenge as existing approaches fail to extract causality with reasonable performance. [**Question/problem:**] We understand causality extraction from requirements as a two-step problem: First, we need to detect if requirements have causal properties or not. Second, we need to understand and extract their causal relations. At present, though, we lack knowledge about the form and complexity of causality in requirements, which is necessary to develop a suitable approach addressing these two problems. [**Principal ideas/results:**] We conduct an exploratory case study with 14,983 sentences from 53 requirements documents originating from 18 different domains and shed light on the form and complexity of causality in requirements. Based on our findings, we develop a tool-supported approach for causality detection (CiRA, standing for Causality in Requirement Artifacts). This constitutes a first step towards causality extraction from NL requirements. [**Contribution:**] We report on a case study and the resulting tool-supported approach for causality detection in requirements. Our case study corroborates, among other things, that causality is, in fact, a widely used linguistic pattern to describe system behavior, as about a third of the analyzed sentences are causal. We further demonstrate that our tool CiRA achieves a macro-F_1 score of 82% on real word data and that it outperforms related approaches with an average gain of 11.06% in macro-Recall and 11.43% in macro-Precision. Finally, we disclose our open data sets as well as our tool to foster the discourse on the automatic detection of causality in the RE community.

© Springer Nature Switzerland AG 2021
F. Dalpiaz and P. Spoletini (Eds.): REFSQ 2021, LNCS 12685, pp. 19–36, 2021.
https://doi.org/10.1007/978-3-030-73128-1_2

Keywords: Causality · Case study · Requirements engineering · Natural Language Processing

1   Introduction

System behavior is usually described by causal relations, e.g. "A confirmation message shall be shown if the system has successfully processed the data." Hence, causal relations are often inherently embedded in the textual descriptions of requirements. Understanding and extracting these causal relations offers great potential for Requirements Engineering (RE); for instance, by supporting the automated derivation of test cases and by facilitating reasoning about dependencies between requirements [7]. However, automated causality extraction from requirements is still challenging for two reasons. First, requirements are mostly expressed by unrestricted natural language (NL) so that the system behavior is specified in arbitrarily complex ways. Second, causality can occur in different forms [2] such as *marked/unmarked* or *explicit/implicit* which makes it difficult to identify and extract the causes and effects. Existing approaches [1] fail to extract causality from NL with a performance that allows for use in practice. Therefore, we argue for the need of a novel method for the extraction of causality from requirements. We understand causality extraction as a two-step problem: We first need to detect whether requirements contain causal relations. Second, if they contain causal relations, we need to understand and extract them. To address both problems, we have to comprehend in which form and complexity requirements causality occurs in practice. This enables us to develop efficient approaches for the automated identification and extraction of causal relations. However, empirical evidence on causality in requirements is presently still weak. In this paper, we report on how we addressed this research gap and make the following contributions (C):

- **C 1:** We report on an exploratory case study where we analyze form and complexity of causality in requirements based on 14,983 sentences emerging from 53 requirement documents. These documents originate from 18 different domains. We corroborate, for example, that causality tends to occur, in fact, in *explicit* and *marked* form, and that about 28% of the analyzed sentences contain causal knowledge about the expected system behavior. This strengthens our confidence in the relevance of our approach.
- **C 2:** We present our tool-supported approach named CiRA (**C**ausality detection **i**n **R**equirement **A**rtifacts), which forms a first step towards causality extraction from NL requirements. We train and empirically evaluate CiRA using the pre-analyzed data set and achieve an macro-F_1 score of 82%. Compared to baseline systems that classify causality based on the presence of certain cue phrases, or shallow ML models, CiRA leads to an average performance gain of 11.43% in macro-Precision and 11.06% in macro-Recall.

– **C 3**: To strengthen transparency and facilitate replication, we disclose our tool, code, and data set used in the case study.[1]

2 Terminology

Causality represents a semantic relation that has been studied by various disciplines, e.g. by psychology [27]. Before we can investigate in which form causality occurs in requirements, we must first understand what causality actually means.

Concept of Causality. Causality is a relation between two events: a causing event (the *cause*) and a caused event (the *effect*). An event is "any situation (including a process or state) that happens or occurs either instantaneously (punctual) or during a period of time (durative)" [19]. The connection between causes and effects is counterfactual [17]: If a cause c_1 did not occur, then an effect e_1 could not have occurred either. Consequently, a causal relation requires that the effect may only occur *if and only if* the cause has occurred. Therefore, in the view of Boolean algebra, a causal relation can be interpreted as an equivalence between a cause and effect ($c_1 \iff e_1$). If the cause is true, the effect is true and if the cause is false, the effect is also false. The relation between a cause and effect can be defined in three different ways [26]: as a *cause*, *enable* or *prevent* relationship.

– c_1 **causes** e_1: If c_1 occurs, e_1 also occurs ($c_1 \iff e_1$). This can be illustrated by REQ 1: "After the user enters a wrong password, a warning window shall be shown." In this case, the wrong input is the trigger to display the window.
– c_1 **enables** e_1: If c_1 does not occur, e_1 does not occur either (e_1 is not enabled). REQ 2: "As long as you are a student, you are allowed to use the sport facilities of the university ($c_1 \iff e_1$)." Only the student status enables to do sports on campus.
– c_1 **prevents** e_1: If c_1 occurs, e_1 does not occur ($c_1 \iff \neg e_1$). REQ 3: "Data redundancy is required to prevent a single failure from causing the loss of collected data." There will be no data loss due to data redundancy.

Temporal Ordering of Causes and Effects. Causes and effects can occur in three different temporal relations [19]. In the first temporal relation, the cause occurs before the effect (*before* relation). REQ 1 requires the user to enter a wrong password before the warning window will be displayed. In this example, the cause and effect represent two punctual events. In the second temporal relation, the occurrence of the cause and effect overlaps: "The fire is burning down the house." In this case, the occurrence of the emerging fire overlaps with the occurrence of the increasingly brittle house (*overlaps* relation). In the third temporal relation (*during* relation), cause and effect occur simultaneously. REQ 2 describes such a relation, as the effect that you are allowed to do sports on the campus is only valid as long as you have the student status. The start and end time of the cause is therefore also the start and end of the effect. Here, both events are durative.

---

[1] A demo of CiRA can be accessed at http://cira.diptsrv003.bth.se/. Our code and annotated data sets can be found at https://github.com/fischJan/CiRA.

Forms of Causality. Causality can be expressed in different forms [2]: *marked* and *unmarked* causality, *explicit* and *implicit* causality, and *ambiguous* and *non-ambiguous* cue phrases.

- **Marked and unmarked**: A causal relation is *marked* if a certain cue phrase indicates causality. The requirement "If the user presses the button, a window appears" is *marked* by the cue phrase "if", while "The user has no admin rights. He cannot open the folder." is *unmarked*.
- **Explicit and implicit**: An *explicit* causal relation provides information about both the cause and effect. The requirement "In case of an error, the systems prints an error message to the console" is *explicit* as it contains the cause (error) and effect (error message). "A parent process kills a child process" is *implicit* because the effect that the child process is terminated is not explicitly stated.
- **Ambiguous and non-ambiguous cue phrases**: Given the difference between *marked* and *unmarked* causality, it seems feasible to deduce the presence of causality in a sentence from the occurrence of certain cue phrases. However, there are cue phrases (e.g. since) that may indicate causality, but also occur in other contexts (e.g. to denote time constraints). Such cue phrases are called *ambiguous*, while cue phrases (e.g. because) that mostly indicate causality are called *non-ambiguous*.

Complexity of Causality. Our previous explanations refer to the simplest case where the causal relation consists of a single cause and effect. With increasing system complexity, however, the expected system behaviour is described by multiple causes and effects that are connected to each other. They are linked either by conjunctions ($c_1 \wedge c_2 \wedge \ldots \iff e_1$) or disjunctions ($c_1 \vee c_2 \vee \ldots \iff e_1$) or a combination of both which increases the complexity of the causal relation. Furthermore, causal relations can not only be contained in a single sentence, but also span over multiple sentences, which is a significant challenge for causality extraction. Additionally, the complexity increases when several causal relations are linked together, i.e. if the effect of a relation r_1 represents a cause in another relation r_2. We define such causal relations, where r_2 is dependent on r_1, as *event chains* (e.g. $r_1 : c_1 \iff e_1$ and $r_2 : e_1 \iff e_2$).

3   Case Study: Causality in Requirement Documents

The case study was performed according to the guidelines of Runeson and Höst [23]. Based on the classification of Robson [22], our case study is exploratory as we seek for new insights into causality in requirement documents. In this section, we describe our *research questions, study objects, study design, study results,* and *threats to validity.* We also give an overview of the *implications of the study* on the causality detection and extraction from requirements.

3.1   Research Questions

We are interested in the form and complexity of causality in requirement documents. Based on the terminology introduced in Sect. 2, we investigate the following research questions (RQ):

- **RQ 1**: To which degree does causality occur in requirement documents?
- **RQ 2**: How often do the relations *cause, enable* and *prevent* occur?
- **RQ 3**: How often do the temporal relations *before, overlap* and *during* occur?
- **RQ 4**: In which form does causality occur in requirement documents?
  RQ 4a: How often does *marked* and *unmarked* causality occur?
  RQ 4b: How often does *explicit* and *implicit* causality occur?
  RQ 4c: Which causal cue phrases are used? Are they mainly *ambiguous* or *non-ambiguous*?
- **RQ 5**: At which complexity does causality occur in requirement documents?
  RQ 5a: How often do multiple causes occur?
  RQ 5b: How often do multiple effects occur?
  RQ 5c: How often does two sentence causality occur?
  RQ 5d: How often do *event chains* occur?

3.2   Study Objects

We considered three criteria when selecting a suitable data set for our case study: 1) the data set shall contain requirements documents that are/were used in practice, 2) the data set shall not be domain-specific, rather it shall contain documents from different domains, and 3) the documents shall originate from different years. Consequently, our analysis is not restricted to a single year or domain, but rather allows for a comprehensive view on causality in requirements. Based on these criteria, we selected the data set provided by Fischbach et al. [7]. To the best of our knowledge, this data set is currently the most extensive collection of requirements available in the RE community. It contains 463 documents, from which the authors extracted and pre-processed 212k sentences. For our analysis, we have randomly selected 53 documents from the data set. Our final data set consists of 14,983 sentences from 18 different domains (see Fig. 1).

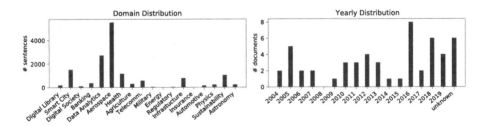

Fig. 1. Descriptive statistics of our data set. The left graph shows the number of sentences per domain. The right graph depicts the year of creation per document.

3.3   Study Design

Model the phenomenon. In order to answer our RQ, we need to annotate the sentences in our data set with respect to certain categories (e.g. *explicit* or *implicit* causality). According to Pustejovsky and Stubbs [21], the first step in each annotation process is to "model the phenomenon" that needs to be annotated. Specifically, it should be defined as a model M that consists of a vocabulary T, the relations R between the terms as well as the interpretations I of terms. RQ 1 can be understood as a binary annotation problem, which can be modeled as:

- **T**: {sentence, causal, not causal}
- **R**: {sentence :: = causal|not causal}
- **I**: {causal = "A sentence is causal if it contains a relation between at least two events, where e1 causes the occurrence of e2", ¬causal = "A sentence is not causal if it describes a state that is independent on any events"}

Modeling an annotation problem has two advantages: It contributes to a clear definition of the research problem and can be used as a guide for the annotators to explain the meaning of the labels. We have modeled each RQ and discussed it with the annotators. In addition to interpretation *I*, we have also provided an example for each label to avoid misunderstandings. After modeling all RQs, the following nine categories emerged, according to which we annotated our data set: Causality , Explicit , Marked , Single Sentence , Single Cause , Single Effect , Event Chain , Relationship and Temporality .

Annotation Environment. We developed our own annotation platform tailored to our research questions.[2] Contrary to other annotation platforms [20] which only show single sentences to the annotators, we also show the predecessor and successor of each sentence. This is required to determine whether the causality extends over one sentence or across multiple ones (see RQ 5c). For the binary annotation problems (see RQ 1, RQ 4a, RQ 4b, RQ 5a–d), we provide two labels for each category. Cue phrases present in the sentence can either be selected by the annotator from a list of already labeled cue phrases or new cue phrases can be added using a text input field (see RQ 4c). Since RQ 2 and RQ 3 are ternary annotation problems, the platform provides three labels for these categories.

Annotation Guideline. Prior to the labeling process, we conducted a workshop with all annotators to ensure a common understanding of causality. The results of the workshop were recorded in the form of an annotation guideline. All annotators were instructed to observe the following annotation rules: First, you should not just check for cue phrases and label the sentence directly as causal, but rather read the sentence completely before making a labeling decision. Otherwise, too many False Positives will be introduced. Second, you should check if the cause is really necessary for the effect to occur. Only if the cause is mandatory for the effect, it is a causal relation.

---

[2] The platform can be accessed at http://clabel.diptsrv003.bth.se/suite.

Table 1. Inter-annotator agreement statistics per category. The two categories Relationship and Temporality were jointly labeled by the first and second author and therefore do not require a reliability assessment.

		Causal		Explicit		Marked		Single sentence		Single cause		Single effect		Event chain		Avg.
		0	1	0	1	0	1	0	1	0	1	0	1	0	1	
Confusion	0	2034	193	24	25	1	22	12	8	41	77	63	72	450	27	
Matrix	1	274	499	39	411	12	464	17	462	43	338	46	318	13	9	
Agreement		84.4 %		87.2 %		93.1 %		95.0 %		76.0 %		76.4 %		92.0 %		86.3 %
Cohen's Kappa		0.579		0.358		0.023		0.464		0.261		0.362		0.27		0.331
Gwet's AC1		0.753		0.84		0.926		0.945		0.645		0.625		0.91		0.806

Annotation Validity. To verify the reliability of our annotations, we calculated the inter-annotator agreement. We assigned 3,000 sentences to each annotator, of which 2,500 are unique and 500 overlapping. Based on the overlapping sentences, we calculated the Cohen's Kappa [3] measure to evaluate how well the annotators can make the same annotation decision for a given category. We chose Cohen's Kappa since it is widely used for assessing inter-rater reliability [25]. However, a number of statistical problems are known to exist with this measure [18]. In case of a high imbalance of ratings, Cohen's Kappa is low and indicates poor inter-rater reliability even if there is a high agreement between the raters (Kappa paradox [6]). Thus, Cohen's Kappa is not meaningful in such scenarios. Consequently, studies [28] suggest that Cohen's Kappa should always be reported together with the percentage of agreement and other paradox resistant measures (e.g. Gwet's AC1 measure [10]) in order to make a valid statement about the inter-rater reliability. We involved six annotators in the creation of the corpus and assessed the inter-rater reliability on the basis of 3,000 overlapping sentences, which represents about 20% of the total data set. We calculated all measures (see Table 1) using the cloud-based version of AgreeStat [11]. Cohen's Kappa and Gwet's AC1 can both be interpreted using the taxonomy developed by Landis and Koch [16]: values ≤ 0 as indicating no agreement and 0.01–0.20 as none to slight, 0.21–0.40 as fair, 0.41–0.60 as moderate, 0.61–0.80 as substantial, and 0.81–1.00 as almost perfect agreement. Table 1 demonstrates that the inter-rater agreement of our annotation process is reliable. Across all categories, an average percentage of agreement of 86% was achieved. Except for the categories ⟨Single Cause⟩ and ⟨Single Effect⟩, all categories show a percentage of agreement of at least 84%. We hypothesize that the slightly lower value of 76% for these two categories is caused by the fact that in some cases the annotators interpret the causes and effects with different granularity (e.g., annotators might break some causes and effects down into several sub causes and effects, while some do not). Hence, the annotations differ slightly. The Kappa paradox is particularly evident for the categories ⟨Marked⟩ and ⟨Event Chain⟩. Despite a high agreement of over 90%, Cohen's Kappa yields a very low value, which "paradoxically" suggests almost no or only fair agreement. A more meaningful assessment is provided by Gwet's AC1 as it did not fail in case of prevalence and remains close to the percentage of agreement. Across all categories, the mean value is above 0.8, which indicates a nearly perfect agreement. Therefore, we assess our labeled data set as reliable and suitable for further analysis and the implementation of our causality detection approach.

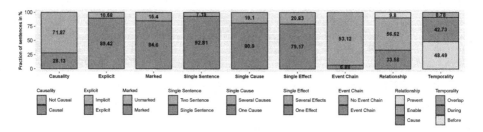

Fig. 2. Annotation results per category. The y axis of the bar plot for the category "Causality" refers to the total number of analyzed sentences. The other bar plots are only related to the causal sentences.

3.4   Study Results

Figure 2 presents the analysis results for each labeled category. When interpreting the values, it is important to note that we analyze entire requirement documents in our study. Consequently, our data set contains records with different contents, which do not necessarily represent all functional requirements. For example, requirement documents also contain non-functional requirements, phrases for content structuring, purpose statements, etc. Hence, the results of our analysis do not only refer to functional requirements but in general to the content of requirement documents.

Answer to RQ1: Figure 2 highlights that causality occurs in requirement documents. About 28% of the analyzed sentences are causal. It can therefore be concluded that causality is a major linguistic element of requirement documents since almost one third of all sentences are causal.

Answer to RQ2: The majority (56%) of causal sentences contained in requirement documents express an *enable* relationship between certain events. Only about 10% of the causal sentences indicate a *prevent* relationship. *Cause* relationships are found in about 34% of the annotated data.

Answer to RQ3: Interestingly, we found that causes and effects occur almost equally often in a *before* and *during* relation. With about 48%, the *before* relation is the most frequent temporal relation in our data set, but only with a difference of about 6% compared to the *during* relation. The *overlap* relation occurred only in a minority (8.78% of the sentences).

Answer to RQ4a: Figure 2 shows that the majority of causal sentences contain one or more cue phrases to indicate the causal relationship between certain events. *Unmarked* causality occurs only in about 15% of the analyzed sentences.

Answer to RQ4b: Most causal sentences are *explicit*, i.e. they contain information about both the cause and the effect. Only about 10% of causal sentences are *implicit*.

Answer to RQ4c: Table 2 provides an overview of the causal cue phrases used in the requirement documents. The left side of the table shows the different cue phrases ordered by word group. On the right side, all verbs used to express causal relations are listed. We order the verbs according to whether they express a *cause*, *enable* or *prevent* relationship. To measure the ambiguity of the individual cue phrases, we introduce the ambiguity factor (AF). We define AF for a cue phrase x as the conditional probability that a sentence is causal given that the cue phrase x occurs in the sentence: $Pr(Causal|X$ is present in sentence$)$. Hence, a high AF value indicates a *non-ambiguous* cue phrase, while low values indicate strongly *ambiguous* cue phrases. Table 2 demonstrates that a number of different cue phrases are used to express causality in requirement documents. Not surprisingly, cue phrases like "if", "because" and "therefore" show AF values of more than 90%. However, there is a variety of cue phrases that indicate causality in some sentences but also occur in other non-causal contexts. This is especially evident in the case of pronouns. Relative sentences can indicate causality, but not in every case, which is reflected by the low AF value. A similar pattern emerges with regard to the used verbs. Only a few verbs (e.g., "leads to, degrade and enhance") show a high AF value. Consequently, the majority of used verbs do not necessarily indicate a causal relation if they are present in a sentence.

Answer to RQ 5a: Figure 2 illustrates that a causal relation in requirement documents often includes only a single cause. Multiple causes occur in only 19.1% of analyzed causal sentences. The exact number of causes was not documented during the annotation process. However, the participating annotators reported consistently that in the case of complex causal relations, two to three causes were usually included. More than three causes were rare.

Answer to RQ5b: Interestingly, the distribution of effects is similar to that of causes. Likewise, single effects occur significantly more often than multiple effects. According to the annotators, the number of effects in case of complex relations is limited to two effects. Three or more effects occur rarely.

Answer to RQ5c: Most causal relations can be found in single sentences. Relations where cause and effect are distributed over several sentences occur only in about 7% of the analyzed data. The annotators reported that most often the cue phrase "therefore" was used to express two-sentence causality.

Answer to RQ5d: Figure 2 shows that *event chains* are rarely used in requirement documents. Most causal sentences contain isolated causal relations and only a few *event chains*.

3.5   Implications for Causality Detection and Extraction

Based on the results of our case study, we draw the following conclusions: Causality matters in requirements documents, which underlines the necessity of an app-

Table 2. Overview of cue phrases used to indicate causality in requirement documents. **Bold** AF values highlight non-ambiguous phrases that mostly indicated causality (Pr(Causal|X is present in sentence) ≥ 0.8).

Type	Phrase	Causal	Not Causal	Ambiguity Factor (AF)	Type	Phrase	Causal	Not Causal	AF
Conjunctions	if	387	41	**0.90**	Cause	force(s/ed)	21	18	0.54
	as	607	1313	0.32		cause(s/ed)	32	10	0.76
	because	78	7	**0.92**		lead(s) to	5	0	**1.00**
	but	100	204	0.33		reduce(s/ed)	48	28	0.63
	in order to	141	33	**0.81**		minimize(s/ed)	28	11	0.72
	so (that)	88	86	0.51		affect(s/ed)	13	19	0.41
	unless	23	4	**0.85**		maximize(s/ed)	11	5	0.69
	while	71	90	0.44		eliminate(s/ed)	8	11	0.42
	once	48	15	0.76		result(s/ed) in	50	43	0.54
	except	9	5	0.64		increase(s/ed)	49	34	0.59
	as long as	12	1	**0.92**		decrease(s/ed)	5	8	0.38
Adverbs	therefore	61	6	**0.91**		impact(s)	37	68	0.35
	when	331	64	**0.84**		degrade(s/ed)	11	2	**0.85**
	whenever	10	0	**1.00**		introduce(s/ed)	11	12	0.48
	hence	21	9	0.70		enforce(s/ed)	2	1	0.67
	where	213	150	0.59		trigger(s/ed)	11	7	0.61
	since	65	32	0.67	Enable	depend(s) on	28	21	0.57
	consequently	2	6	0.25		require(s/ed)	316	262	0.55
	wherever	5	2	0.71		allow(s/ed)	187	130	0.59
	rather	16	30	0.35		need(s/ed)	98	162	0.38
	to this/that end	12	0	**1.00**		necessitate(s/ed)	7	2	0.78
	thus	66	17	**0.80**		facilitate(s/ed)	29	28	0.51
	for this reason	7	3	0.70		enhance(s/ed)	16	4	**0.80**
	due to	91	26	0.78		ensure(s/ed)	145	66	0.69
	thereby	4	2	0.67		achieve(s/ed)	30	24	0.56
	as a result	11	4	0.73		support(s/ed)	128	301	0.30
	for this purpose	1	2	0.33		enable(s/ed)	75	36	0.68
Pronouns	which	277	608	0.31		permit(s/ed)	10	13	0.43
	who	28	52	0.35		rely on	3	5	0.38
	that	732	1178	0.38	Prevent	hinder(s/ed)	1	1	0.50
	whose	16	11	0.59		prevent(s/ed)	38	17	0.69
Adjectives	only	127	126	0.50		avoid(s/ed)	14	23	0.38
	prior to	26	20	0.57					
	imperative	1	3	0.25					
	necessary (to)	36	19	0.65					
Preposition	for	1209	2753	0.31					
	during	327	137	0.70					
	after	133	57	0.70					
	by	506	1171	0.30					
	with	680	1554	0.30					
	in the course of	2	1	0.67					
	through	114	204	0.36					
	as part of	19	51	0.27					
	in this case	18	3	**0.86**					
	before	54	27	0.67					
	until	33	11	0.75					
	upon	25	48	0.34					
	in case of	30	7	**0.81**					
	in both cases	1	0	**1.00**					
	in the event of	15	2	**0.88**					
	in response to	6	7	0.46					
	in the absence of	8	1	**0.89**					

roach for the automatic detection and extraction of causal requirements. The complexity of causal relations ranges from low to medium, since they usually consist of a single cause and effect relationship. However, for the approaches to be applicable in practice, they need to comprehend also more complex relations containing between two to three causes and two effects. Hence, the approaches must be capable of understanding conjunctions, disjunctions and negations in the sentences to fully capture the relationships between causes and effects. Two-sentence causality and *event chains* occur only rarely. Thus, both aspects can initially be neglected in the development of the approaches, while still more than 92% of the analyzed sentences can be covered. Since most causal relations in requirements documents are *explicit*, the detection and extraction of causality is simplified. The information about both causes and effects is embedded directly in the sentences, so that the approaches require little or no *implicit* knowledge. The analysis of the AF values reveals that most of the used cue phrases are ambiguous. Consequently, our methods require a deep understanding of language as causality can not only be deduced from the presence of certain cue phrases but rather from a combination of the syntax and semantics of the sentence.

3.6   Threats to Validity

Internal Validity: A major threat to internal validity are the annotations themselves as an annotation task is to a certain degree subjective. To minimize the bias of the annotators, we performed two mitigation actions: First, we conducted a workshop prior to the annotation process to ensure a common understanding of causality. Second, we assessed the inter-rater agreement by using multiple metrics (Gwet's AC1 etc.). *External Validity*: To achieve reasonable generalizability, we selected requirements documents from different domains and years. As Fig. 1 shows, our data set covers a variety of domains, but the distribution of the sentences is imbalanced. The domains aerospace, data analytics, and smart city account for a large part of the data set (9,724 sentences), while the other 15 domains are underrepresented. Hence, our results do not allow a general conclusion about causality in requirements documents. Future studies should expand to more documents from these underrepresented as well as further domains to achieve a more global insight into causality in requirements documents.

4   Approach: Detecting Causal Requirements

This section presents the implementation of our causal classifier. Initially, we describe our applied methods followed by a report of the results of our experiments, in which we compare the performance of the individual methods.

4.1   Methods

Rule Based Approach. The baseline approach for causality detection involves the use of simple regex expressions. We iterate through all sentences in the test set and check if one of the phrases listed in Table 2 is included. For the positive case, the sentence is classified as causal and vice versa.

Machine Learning Based Approach. As a second approach, we investigate the use of *supervised* ML models that learn to predict causality based on the labeled data set. Specifically, we employ established binary classification algorithms: Naive Bayes (NB), Support Vector Machines (SVM), Random Forest (RF), Decision Tree (DT), Logistic Regression (LR), Ada Boost (AB) and K-Nearest Neighbor (KNN). To determine the best hyperparameters for each binary classifier, we apply Grid Search, which fits the model on every possible combination of hyperparameters and selects the most performant. We use two different methods as word embeddings: Bag of Words (BoW) and Term Frequency-Inverse Document Frequency (TF-IDF). In Table 3 we report the classification results of each algorithm as well as the best combination of hyperparameters.

Deep Learning Based Approach. With the rise of Deep Learning (DL), more and more researchers are using DL models for Natural Language Processing (NLP) tasks. In this context, the Bidirectional Encoder Representations from Transformers (BERT) model [4] is prominent and has already been used for question answering and named entity recognition. BERT is pre-trained on large corpora and can therefore easily be fine tuned for any downstream task without the need for much training data (Transfer Learning). In our paper, we make use of the fine tuning mechanism of BERT and investigate to which extent it can be used for causality detection of requirement sentences. First, we tokenize each sentence. BERT requires input sequences with a fixed length (maximum 512 tokens). Therefore, for sentences that are shorter than this fixed length, padding tokens (PAD) are inserted to adjust all sentences to the same length. Other tokens, such as the classification (CLS) token, are also inserted in order to provide further information of the sentence to the model. CLS is the first token in the sequence and represents the whole sentence (i.e. it is the pooled output of all tokens of a sentence). For our classification task, we mainly use this token because it stores the information of the whole sentence. We feed the pooled information into a single-layer feedforward neural network that uses a softmax layer, which calculates the probability that a sentence is causal or not. We tune BERT in three different ways and investigate their performance:

- **BERT$_{Base}$** In the base variant, the sentences are tokenized as described above and put into the classifier. To choose a suitable fixed length for our input sequences, we analyzed the lengths of the sentences in our data set. Even with a fixed length of 128 tokens we cover more than 97% of the sentences. Sentences containing more tokens are shortened accordingly. Since this is only a small amount, only little information is lost. Thus, we chose a fixed length of 128 tokens instead of the maximum possible 512 tokens to keep BERT's computational requirements to a minimum.

- **BERT$_{POS}$** Studies have shown that the performance of NLP models can be improved by providing explicit prior knowledge of syntactic information to the model [24]. Therefore, we enrich the input sequence with syntactic information and feed it into BERT. More specifically, we add the corresponding Part-of-speech (POS) tag to each token by using the spaCy NLP library [12]. One way to encode the input sequence with the corresponding POS tags is to concatenate each token embedding with a hot encoded vector representing the POS tag. Since the BERT token embeddings are high dimensional, the impact of a single added feature (i.e. the POS tag) would be low. Contrary, we hypothesize that the syntactic information has a higher impact if we annotate the input sentences directly with the POS tags and then put the annotated sentences into BERT. This way of creating linguistically enriched input sequences has already proven to be promising during the development of the NLPL word embeddings [5]. Figure 3 shows how we incorporated the POS tags into the input sequence. By extending the input sequence, the fixed length for the BERT model has to be adapted accordingly. After a further analysis, a length of 384 tokens proved to be reasonable.

- **BERT$_{DEP}$** Similar to the previous fine-tuning approach, we follow the idea of enriching the input sequence by linguistic features. Instead of using the POS tags, we use the dependency (DEP) tags (see Fig. 3) of each token. Thus, we provide knowledge about the grammatical structure of the sentence to the classifier. We hypothesize that this knowledge has a positive effect on the model performance, as a causal relation is a specific grammatical structure (e.g. it often contains an adverbial clause) and the classifier can learn causal specific patterns in the grammatical structure of the training instances. The fixed token length was also increased to 384 tokens.

BERT$_{Base}$: If the process fails, an error message is shown.

BERT$_{POS}$: If **SCONJ** the **DET** process **NOUN** fails **VERB** , **PUNCT** an **DET** error **NOUN** message **NOUN** is **AUX** shown **VERB** . **PUNCT**

BERT$_{DEP}$: If **mark** the **det** process **nsubj** fails **advcl** , **punct** an **det** error **compound** message **nsubjpass** is **auxpass** shown **ROOT** . **punct**

Fig. 3. Input sequences used for our different BERT fine tuning models. POS tags are marked orange and DEP tags are marked blue. (Color figure online)

4.2   Evaluation Procedure

Our labeled data set is imbalanced as only 28.1% are positive samples. To avoid the class imbalance problem, we apply Random Under Sampling (see Fig. 4). We randomly select sentences from the majority class and exclude them from the data set until a balanced distribution is achieved. Our final data set consists of 8,430 sentences of which 4,215 are equally causal and non-causal. We follow the idea of Cross Validation and divide the data set in a training, validation and test set. The training set is used for fitting the algorithm while the validation set is used to tune its parameters. The test set is utilized for the evaluation of the algorithm based on real world unseen data. We opt for a 10-fold Cross Validation as a number of studies have shown that a model that has been trained this way demonstrates low bias and variance [13]. We use standard metrics, for evaluating our approaches: Accuracy, Precision, Recall and F_1 score [13]. When interpreting the metrics, it is important to consider which misclassification (False Negative or False Positive) matters most resp. causes the highest costs. Since causality detection is supposed to be the first step towards automatic causality extraction, we favor Recall over Precision. A high Recall corresponds to a greater degree of automation of causality extraction, because it is easier for users to discard False Positives then to manually detect False Negatives. Consequently, we seek high Recall to minimize the risk of missed causal sentences and acceptable Precision to ensure that users are not overwhelmed by False Positives.

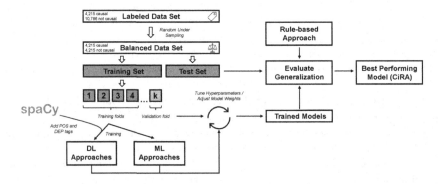

Fig. 4. Implementation and evaluation procedure of our binary classifier

4.3   Experimental Results

Table 3 demonstrates the inability of the baseline approach to distinguish between causal (F_1 score: 66%) and non-casual (F_1 score: 64%) sentences. This coincides with our observation from the case study that searching for cue phrases is not suitable for causality detection. In comparison, most ML based approaches (except KN and DT) show a better performance. The best performance in this category is achieved by RF with an Accuracy of 78% (gain of 13% compared to baseline approach). The overall best classification results are achieved by our DL based approaches. All three variants were trained with the hyperparameters recommended by Devlin et al. [4]. Even the vanilla **BERT$_{Base}$** model shows a great performance in both classes (F_1 score $\geq$ 80% for causal and non-causal). Interestingly, enriching the input sequences with syntactic information did not result in a significant performance boost. **BERT$_{POS}$** even has a slightly worse

Accuracy value of 78% (difference of 2% compared to **BERT**$_{\text{Base}}$). An improvement of the performance can be observed in the case of **BERT**$_{\text{DEP}}$, which has the best F$_1$ score for both classes among all the other approaches and also achieves the highest Accuracy value of 82%. Compared to the rule based and ML based approaches, **BERT**$_{\text{DEP}}$ yields an average gain of 11.06% in macro-Recall and 11.43% in macro-Precision. Interesting is a comparison with **BERT**$_{\text{Base}}$. **BERT**$_{\text{DEP}}$ shows better values across all metrics, but the difference is only marginal. This indicates that **BERT**$_{\text{Base}}$ already has a deep language understanding due to its pre-training and therefore can be tuned well for causality detection without much further input. However, over all five runs, the use of the DEP tags shows a small but not negligible performance gain - especially regarding our main decision criterion: the Recall value (85% for causal and 79% for non-causal). Therefore, we choose **BERT**$_{\text{DEP}}$ as our final approach (CiRA).

Table 3. Recall, Precision, F$_1$ scores (per class) and accuracy. We report the averaged scores over five repetitions and highlight in **bold** the best results for each metric.

		Best hyperparameters	Causal (Support: 435)			Not causal (Support: 408)			
			Recall	Precision	F1	Recall	Precision	F1	Accuracy
Rule based		-	0.65	0.66	0.66	0.65	0.63	0.64	0.65
ML based	NB	alpha: 1, fit_prior: True, embed: BoW	0.71	0.7	0.71	0.68	0.69	0.69	0.7
	SVM	C: 50, gamma: 0.001, kernel: rbf, embed: BoW	0.68	0.8	0.73	0.82	0.71	0.76	0.75
	RF	Criterion: entropy, max_features: auto, n_estimators: 500, embed: BoW	0.72	**0.82**	0.77	**0.84**	0.74	0.79	0.78
	DT	Criterion: gini, max_features: auto, splitter: random, embed: TF-IDF	0.65	0.68	0.66	0.67	0.65	0.66	0.66
	LR	C: 1, solver: liblinear, embed: TF-IDF	0.71	0.78	0.74	0.79	0.72	0.75	0.75
	AB	Algorithm: SAMME.R, n_estimators: 200, embed: BoW	0.67	0.78	0.72	0.8	0.7	0.75	0.74
	KNN	Algorithm: ball_tree, n_neighbors: 20, weights: distance, embed: TF-IDF	0.61	0.68	0.64	0.7	0.63	0.66	0.65
DL based	BERT$_{\text{Base}}$	batch_size: 16, learning_rate: 2e-05,	0.83	0.80	0.82	0.78	0.82	0.80	0.81
	BERT$_{\text{POS}}$	weight_decay: 0.01, optimizer:	0.82	0.76	0.79	0.71	0.83	0.77	0.78
	BERT$_{\text{DEP}}$ (CiRA)	AdamW	**0.85**	0.81	**0.83**	0.79	**0.84**	**0.81**	**0.82**

5   Related Work

As indicated in Sect. 2, many disciplines have already dealt with causality. To the best of our knowledge, we are the first to focus on causality from the perspective of RE. In our previous paper [7], we motivated why the RE community should engage with causality, while in this paper we provide empirical evidence for the relevance of causality in requirement documents and an insight into its form and complexity. Detecting causality in natural language has been investigated by several studies. Multiple papers [14, 29] use handcrafted patterns to identify causal sentences. These approaches are highly dependent on the manually created patterns and show weak performance. Recent papers apply neural networks and exploit, similarly to us, the Transfer Learning capability of BERT [15]. However, we see a number of problems with these papers regarding the realization of our described RE use cases: First, neither the code nor a demo is published, making it difficult to reproduce the results and testing the performance on RE data. Second, they train and evaluate their approaches on strongly unbalanced data sets with causal to non-causal ratios of 1:2 and 1:3, but only report the macro-Recall and macro-Precision values and not the metrics per class. Thus, it is not clear whether the classifier has a bias towards the majority class or not.

6   Conclusion and Next Steps

System behavior is often specified by causal relations in requirements. Extracting this causal knowledge supports automatic test case derivation and reasoning about requirement dependencies [7]. However, existing methods fail to extract causality with reasonable performance [1]. Therefore, we argue for the need of a novel method for causality extraction. We understand causality extraction as a two-step problem: First, we need to detect if requirements have causal properties. Second, we need to comprehend and extract their causal relations. At present, however, we lack knowledge about the form and complexity of causality in requirements, which is needed to develop suitable approaches for these two problems. In this paper, we address this research gap and contribute: (1) an exploratory case study with 14,983 sentences from 53 requirements documents originating from 18 different domains. We found that causality is a widely used linguistic pattern to describe system functionalities and that it mainly occurs in explicit, marked form. (2) CiRA as an approach for the automatic detection of causality in requirements documents. This constitutes a first step towards causality extraction from NL requirements. We empirically evaluate our approach and achieve a macro-F_1 score of 82% on real word data. (3) we disclose our code, tool and annotated data set to facilitate replication.

Two further research directions exist: First, extending the case study and analyzing the sentences from the requirements documents in a more granular way by categorizing them e.g. in functional and non-functional requirements. This would expand our current insight into causality in requirements documents in general by an insight into causality in specific requirement categories. Second, we are enhancing our previous approaches [8,9] to address the second sub-problem: the actual extraction of causal relations.

Acknowledgements. We would like to acknowledge that this work was supported by the KKS foundation through the S.E.R.T. Research Profile project at Blekinge Institute of Technology. Further, we thank Yannick Debes for his valuable feedback.

References

1. Asghar, N.: Automatic extraction of causal relations from natural language texts: A comprehensive survey. arXiv abs/1605.07895 (2016)

2. Blanco, E., Castell, N., Moldovan, D.: Causal relation extraction. In: LREC 2008 (2008)
3. Cohen, J.: A coefficient of agreement for nominal scales. Educ. Psychol. Measur. **20**, 37–46 (1960)
4. Devlin, J., Chang, M.W., Lee, K., Toutanova, K.: BERT: pre-training of deep bidirectional transformers for language understanding. In: NAACL 2019 (2019)
5. Fares, M., Kutuzov, A., Oepen, S., Velldal, E.: Word vectors, reuse, and replicability: Towards a community repository of large-text resources. In: NoDaLiDa 2017 (2017)
6. Feinstein, A.R., Cicchetti, D.V.: High agreement but low Kappa: I the problems of two paradoxes. J. Clin. Epidemiol. **43**, 543–549 (1990)
7. Fischbach, J., Hauptmann, B., Konwitschny, L., Spies, D., Vogelsang, A.: Towards causality extraction from requirements. In: RE 2020 (2020)
8. Fischbach, J., Vogelsang, A., Spies, D., Wehrle, A., Junker, M., Freudenstein, D.: Specmate: automated creation of test cases from acceptance criteria. In: ICST 2020 (2020)
9. Frattini, J., Junker, M., Unterkalmsteiner, M., Mendez, D.: Automatic extraction of cause-effect-relations from requirements artifacts. In: ASE 2020 (2020)
10. Handbook of Inter-rater Reliability: the Definitive Guide to Measuring the Extent of Agreement Among Raters (2012)
11. Gwet, K.: AgreeStat Analytics (Cloud-based version (AgreeStat360) was used in Sept. 2020). https://www.agreestat.com/
12. Honnibal, M., Montani, I.: spaCy NLP library (We use the newest version of the en_core_web_sm model in Sept. 2020). https://spacy.io/
13. James, G., Witten, D., Hastie, T., Tibshirani, R.E.: An Introduction to Statistical Learning. Springer, New York (2013). https://doi.org/10.1007/978-1-4614-7138-7
14. Khoo, C.S.G., Chan, S., Niu, Y.: Extracting causal knowledge from a medical database using graphical patterns. In: ACL 2000 (2000)
15. Kyriakakis, M., Androutsopoulos, I., i Ametllé, J.G., Saudabayev, A.: Transfer learning for causal sentence detection. arXiv abs/1906.07544 (2019)
16. Landis, J.R., Koch, G.G.: The measurement of observer agreement for categorical data. Biometrics **33**, 159–174 (1977)
17. Lewis, D.: Counterfactuals (1973)
18. McHugh, M.L.: Interrater reliability: the Kappa statistic. Biochemia Medica **22**, 276–282 (2012)
19. Mostafazadeh, N., Grealish, A., Chambers, N., Allen, J., Vanderwende, L.: CaTeRS: causal and temporal relation scheme for semantic annotation of event structures. In: EVENTS 2016 (2016)
20. Neves, M., Ševa, J.: An extensive review of tools for manual annotation of documents. Brief. Bioinform. **22**, 146–163 (2019)
21. Pustejovsky, J., Stubbs, A.: Natural Language Annotation for Machine Learning - a Guide to Corpus-Building for Applications. O'Reilly Media Inc., Newton (2012)
22. Robson, C.: Real World Research - A Resource for Social Scientists and Practitioner-Researchers. Wiley, New York (2002)
23. Runeson, P., Höst, M.: Guidelines for conducting and reporting case study research in software engineering. Empir. Softw. Eng. **14**, 131–164 (2009)
24. Sundararaman, D., Subramanian, V., Wang, G., Si, S., Shen, D., Wang, D., Carin, L.: Syntax-infused transformer and bert models for machine translation and natural language understanding (2019)
25. Viera, A., Garrett, J.: Understanding interobserver agreement: the Kappa statistic. Fam. Med. **7**, 360–363 (2005)

26. Wolff, P.: Representing causation. J. Exp. Psychol. Gen. **136**, 82 (2007)
27. Wolff, P., Song, G.: Models of causation and the semantics of causal verbs. Cogn. Psychol. **7**, 276–332 (2003)
28. Wongpakaran, N., Wongpakaran, T., Wedding, D., Gwet, K.: A comparison of Cohen's Kappa and Gwet's AC1 when calculating inter-rater reliability coefficients: a study conducted with personality disorder samples. BMC Med. Res. Methodol. **13**, 61 (2013). https://doi.org/10.1186/1471-2288-13-61
29. Wu, C.H., Yu, L.C., Jang, F.L.: Using semantic dependencies to mine depressive symptoms from consultation records. IEEE Intell. Syst. **20**, 50–58 (2005)

Improving Trace Link Recovery Using Semantic Relation Graphs and Spreading Activation

Aaron Schlutter[1]([✉])[iD] and Andreas Vogelsang[2][iD]

[1] Technische Universität Berlin, Berlin, Germany
aaron.schlutter@tu-berlin.de
[2] Software and Systems Engineering, Universität zu Köln, Köln, Germany
vogelsang@cs.uni-koeln.de
https://www.aset.tu-berlin.de/menue/team/aaron_schlutter/
https://cs.uni-koeln.de/sse

Abstract. [**Context & Motivation**] Trace Link Recovery tries to identify and link related existing requirements with each other to support further engineering tasks. Existing approaches are mainly based on algebraic Information Retrieval or machine-learning. [**Question/ Problem**] Machine-learning approaches usually demand reasonably large and labeled datasets to train. Algebraic Information Retrieval approaches like distance between tf-idf scores also work on smaller datasets without training but are limited in considering the context of semantic statements. [**Principal Ideas/Results**] In this work, we revise our existing Trace Link Recovery approach that is based on an explicit representation of the content of requirements as a semantic relation graph and uses Spreading Activation to answer trace queries over this graph. The approach generates sorted candidate lists and is fully automated including an NLP pipeline to transform unrestricted natural language requirements into a graph and does not require any external knowledge bases or other resources. [**Contribution**] To improve the performance, we take a detailed look at five common datasets and adapt the graph structure and semantic search algorithm. Depending on the selected configuration, the predictive power strongly varies. With the best tested configuration, the approach achieves a mean average precision of 50%, a Lag of 30% and a recall of 90%.

1 Introduction

Trace Link Recovery (TLR) is a common problem in software engineering. While many engineering tasks profit from explicit links between related development artifacts [1, 26], these links are laborious to maintain manually and therefore rarely exist in projects [9]. Automatic TLR approaches aim for supporting engineers in finding related artifacts and creating trace links. Most approaches frame TLR as an Information Retrieval (IR) problem [12]. The IR approach builds upon the assumption that if engineers refer to the same aspects of the system,

© Springer Nature Switzerland AG 2021
F. Dalpiaz and P. Spoletini (Eds.): REFSQ 2021, LNCS 12685, pp. 37–53, 2021.
https://doi.org/10.1007/978-3-030-73128-1_3

similar language is used across different software artifacts. Thus, tools suggest trace links based on Natural Language (NL) content [3].

State-of-the-art approaches use algebraic IR models (e.g., vector space models (VSM), Latent Semantic Indexing (LSI)), or probabilistic models (e.g., Latent Dirichlet Allocation (LDA)) [3]. More recently, machine-learning approaches have also been applied successfully [17]. It is hard to compare the performance of different approaches due to inconsistent use of evaluation metrics and severe threats to validity regarding the used datasets [3]. Algebraic and probabilistic as well as machine-learning approaches rely on implicit models of key terms in requirements (e.g., as points in a vector space or as probability distribution). Trace links are recovered based on similarity notions defined over these models. Therefore, it is hard to analyze and explain *why* specific trace links are identified in the model. Another drawback of machine-learning approaches is the need to train the models on reasonably large datasets. However, TLR datasets usually consists of less than 500 artifacts (at least the ones used in scientific publications [3]). Most are domain-specific, which means that additional care must be taken to ensure that the respective configuration (e.g., the neural net of a machine-learning approach) is not over fitted and thus less reusable for other datasets.

We follow a different approach and base our TLR approach on an explicit model of the knowledge represented in unrestricted NL requirements. Our pipeline translates NL requirements automatically into a semantic relation graph that encodes terms and their relations as vertices and edges. We use Spreading Activation to identify related target requirements (i.e., trace links) for a given query requirement. The semantic search algorithm spreads activation in pulses over the vertices starting from the query vertex. Vertices with higher activation indicate higher relevance for the query. In our previous work [23], we introduced the basic concepts and evaluated the approach with bad to mediocre results.

To improve the results, we take a detailed look at the used datasets with their pitfalls and how this influence the performance of our approach. Subsequently, we elaborate improvements for the graph building and structure as well as resulting adjustments to Spreading Activation. We applied and evaluated the approach on 5 datasets, commonly used in TLR research, in terms of *mean average precision* and *Lag* for a result list length of 5, 10, and 30 trace link candidates. With the best tested configuration, our approach achieves an average precision around 50%, Lag around 30% and recall around 90%.

2   Background

2.1   Trace Link Recovery

Requirements traceability is defined as "the ability to describe and follow the life of a requirement, in both a forwards and backwards direction" [6], i.e., over several phases and periods of refinement during those phases. A trace link states a dependency, relation, or similarity between two artifacts, the source and the target. We do not distinguish the type of links in the following work as we interpret all of them as some kind of relation.

Borg et al. [3] present a mapping study of IR approaches for traceability. They focus on text retrieval and classify 79 publications including their approaches based on the used retrieval model. Borg et al. treat the IR process as essential, NLP techniques are interpreted as an optional prerequisite. They differentiate between algebraic, probabilistic, and statistical language models as well as miscellaneous aspects like weighting scheme, similarity measures/distance functions, and enhancement strategies. The majority of classified publications applied an algebraic model, while most were evaluated in experiments on benchmarks (without human intervention) and used precision and recall as metrics.

Our approach does not match in general any of the retrieval models or their categories as we do not transfer requirements into a mathematical (algebraic, probabilistic, or statistical) model nor do we primarily apply any mathematical operations. We use several NLP techniques to analyze textual requirements, extract terms and their relations [21]. Subsequently, we transfer the results into a semantic relation graph and use a semantic search algorithm to find related artifacts. This includes some of the miscellaneous aspects of Borg et al. like phrasing, term frequency, and optionally similarity measures. Our approach does not support any other kind of requirements than textual ones.

Likewise, we do not focus on automatic linking of related requirements without human intervention since we assume that trace links are usually used in sophisticated scenarios where every link is created manually based on given rules or guidelines. In our view, this cannot be achieved due to the ambiguity of natural language and the variety of guidelines. Instead, we want to support engineers who manually create trace links with a sorted candidate list.

2.2 Knowledge Representation

Knowledge representation focuses on the depiction of information that enables computers to solve complex problems. Borgida et al. [4] already noted in 1985 that knowledge representation is the basis for requirements engineering.

Dermeval et al. [5] report on the use of ontologies in requirements engineering in their systematic literature review. They reviewed 67 publications from academic and industrial application contexts dealing with different types of requirements. While only 34% reused existing ontologies, most of them specified their own ontology. The largest number of publications rely on textual requirements as the RE modeling style, especially in the specification phase.

Robeer et al. [19] automatically derive conceptual models from user stories. The models enable discussion between stakeholders and show promising accuracy results (precision and recall between 80–92%). They use heuristics to analyze the user stories due to semi-structured natural language.

In a former publication [21], we presented an NLP pipeline that extracts knowledge from requirement documents and transforms it into a graph representing RDF[1] triples (subjects and objects become vertices, predicates become edges). The two generated sample graphs were not well-connected and yielded only a subset of fully connected vertices in a main graph. They used one graph to show the separation of two subsystems in an exemplary requirement specification.

2.3   Existing Approach

In our previous work [23], we used NLP techniques to extract semantic information from requirements and build a semantic relation graph as a knowledge base. We used Spreading Activation as semantic search algorithm to find, for a given query (trace link source), all related information (targets). The approach is fully automated to enable engineers to apply TLR to NL requirements without any further effort (e.g., maintain external resources like dictionaries). Comparable to machine-learning, an appropriate configuration must be initially determined for the approach. Such a configuration includes the general structure of the semantic relation graph as well as the configuration of Spreading Activation.

We built an NLP pipeline consisting of Stanford CoreNLP [15] and Deep-SRL [8] to extract information from the requirements and to build the graph. The main goal while building the semantic relation graph is to depict semantic parts of common NL in vertices and connect these with each other based on their relation. The graph structure resembled a tree structure with multiple roots (identifier vertices) and vertices arranged in levels for verbs, arguments, and noun phrases. This structure supported that short phrases have a greater distance (i.e., are less relevant) to a certain specification than complex phrases or whole (verb) statements [22]. We used the given structure of semantic role labeling (SRL), which associates arguments with a semantic role to their predicate within a sentence. In addition, we used several basics like Lemmatizing or part-of-speech tagging to split and compare natural language phrases or filter insignificant parts.

To answer trace queries with the help of the semantic relation graph, we used Spreading Activation. We utilized the state-of-the-art options and algorithm by Hartig [7] to configure the spreading of activation[2]. There are several modes and parameter to adjust attenuation, sending or branching to calculate the activation values during the pulsation. To find an appropriate configuration, we used an exploratory approach and discovered 500 random configurations. The final activation values of the identifier vertices are used to build a sorted candidate list with all (reachable) targets for a query based on their relation.

We evaluated the approach on five datasets of Huffman Hayes et al. [13]. We achieved bad to mediocre results, i.e., the top configurations have a mean average precision of 40% and a Lag of 50%, while most configurations have much worse results. Certain discrete modes are clearly favored and the chosen values for the numerical parameters seem to fit as there are (local) maxima identifiable. To

---

[1] https://www.w3.org/TR/2014/REC-rdf11-concepts-20140225/.

[2] Our Implementation: https://github.com/tub-aset/spreadingactivation.

compare the performance of configurations, we ranked them and found out that a configuration has a constant quality for different datasets. This is probably the case because the basic (syntactic and semantic) structure of requirements or natural language per se is usually similar, even if the concrete content differs.

We concluded that an almost optimal configuration, once determined, may be reused on other datasets with different requirements, key terms and relations. Even if there are specific configurations for individual datasets that perform better in individual case, the evaluation has shown that a uniform configuration for all datasets provides almost equivalent results. That is probably because the basic characteristics of natural language (like the relation between semantic arguments) and trace links (e.g., related requirements use equivalent terms) are similar in different domains. Furthermore, we concluded that improvements for the approach will mainly be found in adjustments of the graph.

Table 1. Vocabulary of all Datasets

	Infusion P.		CCHIT		GANNT		CM-1		WARC		WARCf	
	High	Low	High	Low	High	Low	High	Low	High	Low	High	Low
Stopwords	1,090	486	875	9,621	228	635	1,954	6,874	355	512	404	654
Token	2,325	948	1,560	16,311	327	940	3,159	10,310	634	987	745	1,239
Token (SRL)	2,036	813	1,552	16,245	275	936	3,030	9,502	616	947	741	1,230
Sentences	296	52	121	1,167	29	107	276	1,019	66	98	66	99
Sent. (SRL)	186	49	120	1,163	29	106	274	811	60	85	66	99
Noun phrases	1,251	527	908	9,921	215	595	1,865	6,150	369	538	417	668
NP (SRL)	961	382	878	9,706	170	578	1,715	5,139	343	479	408	640
Lemma	2,325	948	1,560	16,311	327	940	3,159	10,310	634	987	745	1,239
(unique)	646	225	563	1,978	125	196	696	1,718	262	426	287	462
Representatives	46	45	15	157	6	29	42	360	8	15	8	16
Nominal	30	42	12	105	2	21	20	283	1	8	1	8
Pronominal	22	10	5	73	4	17	17	87	4	7	4	8
Proper	0	11	0	7	0	1	9	202	3	3	3	3

3   Datasets: Characteristics of Requirements

We use five common datasets of Huffman Hayes et al. [13] to evaluate our approach: Infusion Pump, CCHIT-2-WorldVista, GANNT, CM-1, and WARC. They come from different domains like health care, science, or business, and are different according to their size and scope of the requirements. We also used them in our previous work [23]. In the meantime we found out that the download link[3] of Huffman Hayes et al. contains a faulty WARC dataset where several requirements missed parts of their content (i.e., the original dataset contains the identifier of a requirement as prefix followed by a dash but Huffman Hayes et al. incorrectly removed text before the last rather than the first occurrence of a dash). To circumvent this issue, we use a fixed version[4], hereby called WARCf.

---

[3] https://selab.netlab.uky.edu/AIRE-2019-hayes-payne-leppelmeier-meta-data.zip.

[4] http://sarec.nd.edu/coest/datasets/WARC.zip.

Table 1 contains a statistical overview of the vocabulary of all datasets, separated into high-level (query) and low-level (target) requirements. At first, we removed all stop words using the NLTK[5] stop word list. The suffix (SRL) for tokens means that there is an SRL argument that covers this token in terms of indices. For Lemma of tokens we also determined how many unique occurrences there are in order to draw conclusions about the reuse of words. To parse the natural language text, we use Stanford CoreNLP [15] and DeepSRL [8].

Table 1 reveals that DeepSRL does not consider all token and even miss several sentences (especially in Infusion Pump high and CM-1 low). These tokens and sentences are not parsable, e.g., the sentence does not contain an (identified) verb. The same observation also applies to noun phrases. On average, only 90% of all noun phrases are covered by an SRL argument, with a minimum of almost 72%. It is also evident that the lemma of tokens is often reused with an average of about 4 times to a maximum of more than 8 times.

Figure 1 and 2 show the intersection of unique lemma vocabulary for all datasets. While the lemma of tokens is much more reused between high-level and low-level requirements, this is not the case for noun phrases.

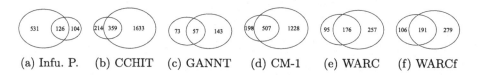

(a) Infu. P.   (b) CCHIT   (c) GANNT   (d) CM-1   (e) WARC   (f) WARCf

Fig. 1. Intersection of Token lemmas

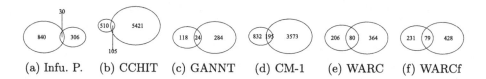

(a) Infu. P.   (b) CCHIT   (c) GANNT   (d) CM-1   (e) WARC   (f) WARCf

Fig. 2. Intersection of Noun phrase lemmas

4   Approach Revisions

Based on the insights in Sect. 3, we revise our approach to improve the performance on TLR. The analysis of the datasets shows that we need to focus more on single words than on whole phrases and also take the frequency into account. Furthermore, we need to consider parts that are not parsed by SRL.

---

[5] https://www.nltk.org.

4.1   Knowledge Base Construction

We reuse the NLP pipeline [23] with the same techniques such as part-of-speech tagging, lemmatizing (morphological analysis), dependency parser (grammatical structure), coreference resolution and SRL. In contrast to our previous work, we changed significantly the building and the resulting structure of the graph.

Unlike before we choose a bottom-up approach, starting with single words. While we continue to believe that words like adjectives, verbs or adverbs without their context or role should not be used, all nouns (identified by their POS tag) are added as a vertex to the graph. On the next level, nouns are assembled into phrases. Therefore, all noun and person phrases (NP and PP within the dependency tree) that do not contain a verb phrase (VP) are also added as vertices to the graph and connected via edges to their contained nouns. If nouns or phrases occur multiple times, the same vertices are reused. To support this reuse, the lemma of words is used. In addition, phrases are trimmed based on the POS tag, e.g., determiners or prepositions at the beginning are removed.

Coreferences are also applied to the graph. If a phrase is pronominal, no vertex is added but instead the vertex of the representative is used. Other coreferences are depicted as edges between the reference and representative vertices.

Last but not least verb vertices are added to graph. As before, we assume that a predicate describe a larger context between their arguments. This time, we do not add separate argument vertices but instead use directly the noun and phrase as argument vertices. Arguments are assigned to their verb in two ways. If there is an SRL predicate with an argument, that covers a phrase or a single noun, we add the verb as vertex and connect it via an argument edge labeled with the semantic role to the corresponding vertex of the noun or phrase. But if SRL was not able to find a predicate for an argument, the dependency tree is used to find a superordinated verb phrase (VP). In this case, the verb of the verb phrase is added as verb vertex and linked to the argument vertex without a certain semantic role label, i.e., we just use *arg* as any argument.

Verb vertices are only reused, if the lemma of the verb and its arguments are equal. The arguments are differentiated according to their type. If the verb has any noun or phrase argument, only they are used to check the equality of verbs. If no such argument is present, all other verb arguments are taken into account.

Adjectives and adverbs are also represented in the graph. While adjectives are implicit contained in phrases, adverbs are explicitly added as parts to the label of verb vertices. Again, if they are covered by an SRL argument, we use additional to the lemma itself the semantic role as prefix. If an adverb is not covered by SRL, we use the same process as for phrase arguments to search within the dependency tree for the corresponding verb.

To omit the distance between identifier and short phrase vertices, we eliminated the identifier vertices. Instead, the requirement identifiers are added as metadata to all vertices of a requirement. If a noun or phrase is used several times within a requirement, the identifier is also indicated several times. Thus, the frequency can be considered in the further course.

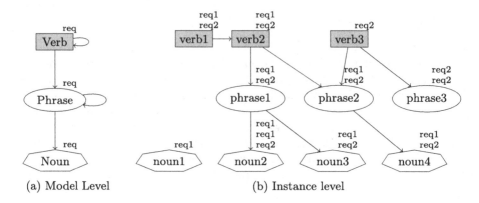

(a) Model Level   (b) Instance level

Fig. 3. Graph structure

Figure 3 shows the model and an abstract instance of the graph structure. While most parts are used by both requirements *req1* and *req2*, therefore indicating a high relevance to each other, there are some vertices to which this does not apply. Vertex *noun1* is only used once in requirement *req1*, likewise *phrase3* is used twice only in *req2*. Although *noun1* has no relation to the other words, its vertex can be attributed to the corresponding requirement and may contribute to TLR, if in the further course more requirements are added that will also use it.

In contrast to our previous graphs, even if there is no verb within a sentence, other (relevant) parts like phrases or nouns are assigned to requirements and therefore taken into account for TLR. Adjectives and adverbs are closer the words they refer to. If SRL is not able to parse the sentence or single phrases, we use dependency relations to find the corresponding verb relation.

4.2   Semantic Search

The graph algorithm Spreading Activation consists of three phases. In the initial phase, the start vertices are activated, i.e., they will be assigned an initial activation value. During the spreading phase, this activation is step wise distributed over the graph, i.e., the activation of a vertex is transferred to related (connected) vertices. These steps are called pulses and at the end of each pulse, a termination condition is checked to stop the pulsation. In the final phase, a sorted candidate list is created using the activation values to sort all vertices by relevance.

Due to the fact that there are no identifier vertices for requirements anymore, we need to change phase 1 and 3 of Spreading Activation. For the initial activation, we use the metadata by selecting the vertices that are included in the query requirement. To consider the frequency, vertices with multiple query occurrences are higher activated, i.e., we use tf-idf values [14] as initial activation values. The tf-idf values are also used in phase 3, where the target requirements are determined by adding the final activation values with tf-idf as factor.

Table 2. Semantic relation graphs

Dataset	Vertices			Edges		
	High	Both	Low	High	Both	Low
Infusion P.	1,130	80	306	1,249	1,059	352
CCHIT	655	238	5,552	582	5,137	6,894
GANNT	156	34	431	141	378	564
CM-1	1,096	378	4,514	1,094	5,765	5,120
WARCf	294	149	591	252	803	582

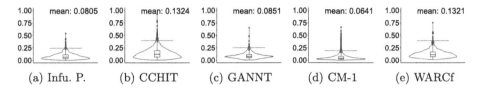

(a) Infu. P.   (b) CCHIT   (c) GANNT   (d) CM-1   (e) WARCf

Fig. 4. Distribution of tf-idf values

In addition, we also adjust phase 2, the spreading process. We now consider the level/type of vertices and edges, e.g., a noun vertex might need a different treatment than a verb one or an *A1* edge might be more relevant than an adjunct one. Therefore, we use different modes and parameter based on the type, except for p_{max} as a global parameter. In case of the edge weight, we introduce a new parameter W that replaces the CONSTANT factor of 1 (cf. [23, Table IV]) to give various semantic roles different weight. To optimally map the relevance of semantic roles to W, we follow the formula in the PropBank Guidelines: A0 > A1 > A2-A6 > AM [2, p. 8]. For the argument type *arg*, we use the *A1* weight.

5   Evaluation

We evaluate the approach for TLR on the five given datasets (Infusion Pump, CCHIT-2-WorldVista, GANNT and CM-1 of [13] as well as WARCf).

We build semantic relations graphs for each dataset, containing both high and low level requirement specifications. Table 2 gives an overview of the number of vertices and edges that have only high-level or low-level requirements as origin and intersecting elements (comparable to the Venn diagrams in Fig. 1 and 2).

To find an appropriate Spreading Activation configuration, we used an exploratory approach and discovered 500 random but valid configurations. While the modes are randomly chosen, the numerical parameter ranges are $d \in [0-1]$, $\tau \in [0-1]$, $p_{max} \in [1-50]$, and $W \in [0-1]$. A configuration is valid if spreading stops before the given number of pulses for all queries of all datasets. Otherwise, this indicates too strong restriction to spread activation and lead to no results.

As we use tf-idf values for initial activation, we need to adapt τ, the minimum activation needed to spread in current pulse, for each dataset. Figure 4 shows the distribution of tf-idf values for all datasets. While the basic shape is equal for all datasets, they all have their mean at a different level. If we would choose the same τ for all datasets, this would result in different limitations of the tf-idf activation values for the same configurations. Instead, we add per dataset a factor of three times the mean value to τ, indicated as red lines in Fig. 4.

5.1 Metrics

There are several evaluation metrics that depend on different goals when evaluating TLR. Shin et al. [24] performed a systematic literature review and defined three different goals. Goal 1 is to find trace links with high accuracy, e.g., to support tasks like coverage analysis. Goal 2 is to find relevant requirements excluding irrelevant requirements to reduce unnecessary effort for human analysts. Goal 3 is to rank all requirements so that the relevant ones are near the top of the retrieved list, also to reduce human effort. Our approach supports goal 3, as we build a ranked list of all requirements.

To evaluate the achievement of goal 3, Shin et al. mention three different metrics, i.e., average precision (AP), Lag, and AUC (area under the ROC curve). Each metric focuses on different weighting schemes for the position in the ranked list. AP assigns a non-proportionally higher weight to a correct link ranked at the top and thereby rewards correct links at the top. Lag assigns a non-proportionally higher weight to a correct link ranked at the bottom of the list, which penalizes those links. AUC uses the same weight for all correct links but is not applicable as it is a classification accuracy metric and not a rank accuracy metric [10]. While a high value is desired for AP, Lag indicates how many incorrect links are proposed before a correct one and should be as low as possible.

Shin et al. show five different types of thresholds for the ranked list. Despite the fact that they recommend relative thresholds rather than absolute ones, we use ND (number of retrieved requirements) with the values 5, 10, and 30, which cuts the list after a fixed number of retrieved requirements. We justify this decision by assuming that the approach should be used in setups including a lot of requirements, but a user is not capable to check through thousands of potential candidates. A relative threshold such as 10% of the list would yield only 2 results for Infusion Pump but more than 40 for CCHIT.

In addition to AP and Lag, we evaluate the recall [14] at certain ND, i.e., the fraction of relevant links that are found until ND limit is reached. In case there are more valid links than ND, we use ND as maximum of relevant links.

The metric values are summarized on *average* across all traces, i.e., mean average precision (MAP) for AP. We calculated the metrics only for source artifacts that are linked because our approach does no classification and provides results for all queries (even for queries without any correct answer).

5.2   Results for Datasets

Figure 5 shows the MAP, Lag and recall values at ND 5, 10, and 30 for each dataset. The numbers above and below are the respective top values. While the majority of configurations already show improved results, the best ones achieve a MAP roughly around 50% and a Lag around 30% on all ND thresholds, depending on the dataset. While Lag of CCHIT is comparable to the other datasets, the MAP indicates that correct results are not in top results but shown at a lower position.

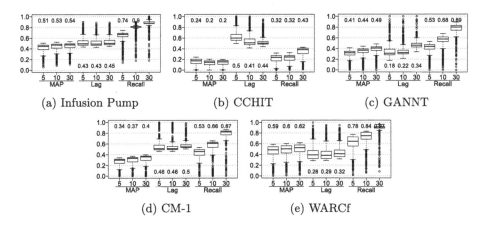

Fig. 5. MAP and Lag at ND 5, 10, and 30

Figure 6 shows all search results for the CM-1 dataset as heat maps. The x-axis lists the 500 configurations, the y-axis contains 155 queries for each linked high-level requirement. The color scheme indicates the recall, i.e., the ratio of correct links in the answer list, from light gray for no correct links at all to dark gray for 100% of all correct links, independent of the position in the result list. The darker a column is, the better a configuration, and the darker a row it is easier to answer a query correctly. There are a few queries where, regardless of the configuration, valid links are found often, e.g., around query 135 and 75. This pattern becomes clearer as the ND threshold increases. In addition, there are some queries where results almost never contain a valid link, e.g., around 140. Comparable to this horizontal patterns, there are vertical ones with some of the best configurations around 300 and some of the worst around 160.

The corresponding heat maps of the other datasets (Fig. 7) have similar vertical patterns at the same places, especially for the configurations around 300. This leads us to the assumption that a configuration has a constant quality for different datasets. We already showed that the quality of configurations is quite constant for different datasets [23].

Table 3 shows the best metric values of all configurations for each dataset and threshold of our new approach compared to our previous approach [23, cf.

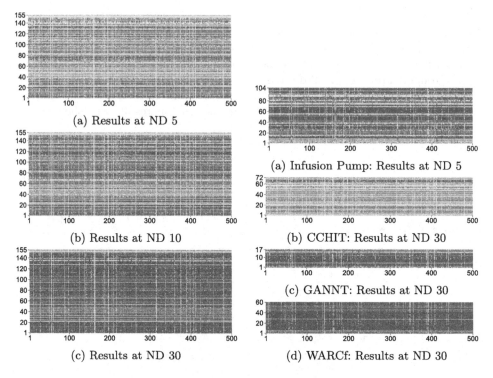

Fig. 6. Results of the CM-1 dataset

(a) Results at ND 5

(b) Results at ND 10

(c) Results at ND 30

(a) Infusion Pump: Results at ND 5

(b) CCHIT: Results at ND 30

(c) GANNT: Results at ND 30

(d) WARCf: Results at ND 30

Fig. 7. Other results at certain ND

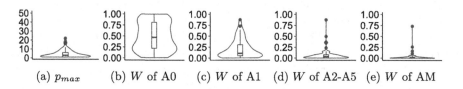

(a) p_{max} (b) W of A0   (c) W of A1   (d) W of A2-A5   (e) W of AM

Fig. 8. Parameter distributions of top-50

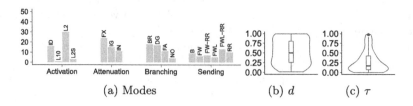

(a) Modes   (b) d (c) τ

Fig. 9. Parameter distributions of top-50 noun vertices

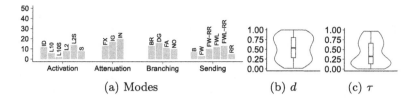

(a) Modes   (b) d (c) τ

Fig. 10. Parameter distributions of top-50 phrase vertices

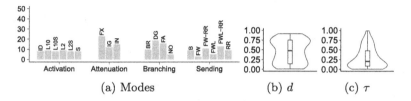

(a) Modes   (b) d (c) τ

Fig. 11. Parameter distributions of top-50 verb vertices

Table 3. Best configurations

ND		Inf. P.			CCHIT			GANNT			CM-1			WARCf		
		5	10	30	5	10	30	5	10	30	5	10	30	5	10	30
Per metric	New approach	MAP	.51	.53	.54	.24	.20	.20	.41	.44	.49	.34	.37	.40	.59	.60	.62
		Lag	.43	.43	.45	.50	.41	.44	.18	.22	.34	.46	.46	.50	.28	.29	.32
		Recall	.74	.90	1.0	.32	.32	.43	.53	.68	.89	.53	.66	.87	.78	.84	.97
	Improvement to [23]	MAP	.08	.08	.07	.13	.10	.10	.06	.05	.07	.08	.09	.09	.26	.24	.23
		Lag	.07	.09	.07	.16	.19	.10	.11	.06	.0	.08	.05	.05	.21	.19	.18
Overall	New approach	MAP	.52	.53	.54	.22	.18	.19	.35	.39	.43	.32	.36	.39	.59	.60	.62
		Lag	.43	.43	.45	.54	.47	.46	.30	.28	.42	.47	.48	.51	.29	.29	.32
		Recall	.73	.84	.92	.29	.29	.41	.47	.61	.79	.47	.63	.84	.75	.83	.92
	Improvement to [23]	MAP	.12	.10	.10	.14	.11	.12	.05	.04	.05	.06	.08	.09	.29	.27	.26
		Lag	.09	.10	.10	.17	.14	.11	.13	.14	.0	.07	.03	.04	.25	.24	.23

TABLE VIII], with the deviation by how much the metric value have improved. The overall best configuration is determined by the best harmonic mean between MAP and Lag. In all cases the new approach has better results in metric values with an average increase of almost 10%. The WARC dataset has improved even more, but this is due to the fact that the old evaluation used the unfixed version. Except for CCHIT, the recall is above 90% at ND30.

5.3   Limitations

In our previous work, we stated that there are mainly two parts of the approach that affect the performance results, the graph and Spreading Activation. As we have already seen in Sect. 3, also the datasets with their characteristics need to be taken into account to have an influence on the performance. Even though we have tried to take a closer look at them and to address or avoid certain problems, there will still be some that have a negative impact.

For example, the datasets reveal at a brief overview that some of the sentences or statements are formulated in a very complex way or contain a list of key points. This may lead to heavily branched (verb) vertices, which might be handled differently by Spreading Activation than other (simpler or more common) sentence structures, e.g., the initial activation might be broadly distributed and lead to super-spreading of those vertices or to no spreading at all due to a high τ.

In addition, we rely on the quality of the NLP tools, especially the complex tasks SRL and dependency trees. If they fail to capture the natural language, these errors will also appear in the graph and probably result in a lower performance.

Figure 8, 9, 10 and 11 show the parameter distribution of top-50 configurations for all levels and types. Figure 8 reveals that the first argument is most important, while adjunct arguments seem to almost be ignored. This insight does not take place in the graph structure, e.g., arguments are considered uniformly to find identical verbs. The parameter τ seems to have a low limit below 0.5 (at least for nouns and verbs), this may be due to the fact that three times the mean tf-idf value as upper limit is too restrictive in combination with a high p_{max}.

Furthermore, we do not merge or connect semantically similar vertices. Two or more words/phrases are semantically similar if they have the same meaning but different syntax. There are different approaches to identify such semantic similarities like word embeddings (*word2vec* [16], *GloVe* [18]) or databases (WordNet[6]). While word embeddings demand additional computation and a reasonably large dataset, databases often contain only general but not domain-specific terms.

Overall, it is difficult to determine the influence of individual factors or their combination, since determining a good and valid configuration for Spreading Activation by random selection requires a lot of computation. It is possible that some of the improvements presented are also counterproductive, but we are not able to verify this in detail. Borg et al. [3] stated that TLR as an IR-problem is often based on empirical software engineering research with small datasets and therefore may conduct in "the cave of IR evaluation".

6   Discussion

We achieve improved results in the evaluation without any explicit assumptions on the requirements (e.g., certain patterns like user stories). This time we included a common IR-solution, the tf-idf values into our approach to improve the quality. The graph and the algorithm scales for different sizes of datasets and is (almost) immediately applicable (i.e., no training needed, only one pass through the NLP pipeline). Also, the graph adapts immediately to changes in

---

[6] https://wordnet.princeton.edu/.

the data as new requirements are parsed and vertices/edges are inserted directly into the graph as well as existing vertices/edges are removed if requirements are removed. Finding an optimal configuration for the semantic search algorithm is a non-trivial task. Certain discrete modes are clearly favored and the chosen values for the numerical parameters seem to fit as there are (local) maxima identifiable in Fig. 8, 9, 10 and 11. Table 3 shows that the performance of the overall best configuration is comparable to configurations with an individual best metric result. We assume that an almost optimal configuration, once determined, may be reused for other datasets as well.

Unfortunately, as we have already shown in our previous work [23], there are currently few publications for comparison of our approach. Most of them do not use the datasets or only report precision, recall, or F1-score. Some approaches also require manual effort, which may be very specific to the dataset in question. For comparison, we used a syntactical VSM approach with tf-idf values for lemmas and compare requirements using cosine similarity [20, 25].

Compared to our previous work [23], the results have improved. While the top values of MAP and Lag only slightly changed, the average performance of a random configuration is greatly increased (cf. Fig. 5d and 6 and [23, Fig. 10d and 11]). This is due to the fact that MAP and Lag have non-proportional weights, which causes strong variation due to changes at the top of the result list, but only small in the further part. In contrast, the syntactical approach still outperforms the semantic approach in terms of MAP and Lag but this time our approach is quite close to baseline performance. From a practical point of view, the recall of about 90% means that an engineer will have seen almost all links in the sorted candidate list after only 30 items, independent of the position within the list. Huffman Hayes et al. [11] rate a recall above 80% as excellent, which we achieve for every dataset except CCHIT with ND30.

7 Conclusion

In this paper, we improve our novel approach for Trace Link Recovery using semantic relations between parts of natural language, stored in a semantic relation graph, and searched by a semantic search algorithm. While the approach is fully automated, it does not have any prerequisites with regard to the format or the content of the natural language (except for English language) and is scalable to various sizes of corpora. We achieve better results than before by incorporating some general characteristics of the datasets and their requirements. In addition, we improved Spreading Activation by using a modified graph structure, tf-idf as activation values and consider even more semantics of natural language.

References

1. Antoniol, G., Canfora, G., Casazza, G., De Lucia, A., Merlo, E.: Recovering traceability links between code and documentation. Trans. Softw. Eng. **28**(10), 970–983 (2002). https://doi.org/10.1109/TSE.2002.1041053

2. Bonial, C., Bonn, J., Conger, K., Hwang, J., Palmer, M., Reese, N.: English PropBank annotation guidelines (2015). https://raw.githubusercontent.com/propbank/propbank-documentation/master/annotation-guidelines/Propbank-Annotation-Guidelines.pdf
3. Borg, M., Runeson, P., Ardö, A.: Recovering from a decade: a systematic mapping of information retrieval approaches to software traceability. Empir. Softw. Eng. **19**(6), 1565–1616 (2013). https://doi.org/10.1007/s10664-013-9255-y
4. Borgida, A., Greenspan, S., Mylopoulos, J.: Knowledge representation as the basis for requirements specifications. In: Brauer, W., Radig, B. (eds.) Wissensbasierte Systeme, vol. 112, pp. 152–169. Springer, Heidelberg (1985). https://doi.org/10.1007/978-3-642-70840-4_13
5. Dermeval, D., et al.: Applications of ontologies in requirements engineering: a systematic review of the literature. Require. Eng. **21**(4), 405–437 (2015). https://doi.org/10.1007/s00766-015-0222-6
6. Gotel, O., Finkelstein, C.W.: An analysis of the requirements traceability problem. Require. Eng. (1994). https://doi.org/10.1109/ICRE.1994.292398
7. Hartig, K.: Entwicklung eines information-retrieval-systems zur Unterstützung von Gefährdungs- und Risikoanalysen. Ph.D. thesis, Technische Universität Berlin (2019). https://doi.org/10.14279/depositonce-8408
8. He, L., Lee, K., Lewis, M., Zettlemoyer, L.: Deep semantic role labeling: what works and what's next. In: Association for Computational Linguistics, pp. 473–483. ACL (2017). https://doi.org/10.18653/v1/p17-1044
9. Heindl, M., Biffl, S.: A case study on value-based requirements tracing. European Software Engineering Conference (2005). https://doi.org/10.1145/1081706.1081717
10. Herlocker, J.L., Konstan, J.A., Terveen, L.G., Riedl, J.T.: Evaluating collaborative filtering recommender systems. Trans. Inf. Syst. **22**(1), 529–565 (2004). https://doi.org/10.1145/963770.963772
11. Huffman Hayes, J., Dekhtyar, A., Sundaram, S.K.: Advancing candidate link generation for requirements tracing: the study of methods. Trans. Softw. Eng. **32**(1), 4–19 (2006). https://doi.org/10.1109/TSE.2006.3
12. Huffman Hayes, J., Dekhtyar, A., Osborne, J.: Improving requirements tracing via information retrieval. In: Requirements Engineering (2003). https://doi.org/10.1109/ICRE.2003.1232745
13. Huffman Hayes, J., Payne, J., Leppelmeier, M.: Toward improved artificial intelligence in requirements engineering: metadata for tracing datasets. In: Artificial Intelligence for Requirements Engineering, pp. 256–262. IEEE (2019). https://doi.org/10.1109/REW.2019.00052
14. Manning, C.D., Raghavan, P., Schutze, H.: Introduction to Information Retrieval. Cambridge University Press, Cambridge (2008). https://doi.org/10.1017/CBO9780511809071
15. Manning, C.D., Surdeanu, M., Bauer, J., Finkel, J., Bethard, S.J., McClosky, D.: The Stanford CoreNLP natural language processing toolkit. In: System Demonstrations, pp. 55–60. ACL (2014)
16. Mikolov, T., Chen, K., Corrado, G., Dean, J.: Efficient estimation of word representations in vector space. Computing Research Repository (2013). https://arxiv.org/abs/1301.3781
17. Mills, C.: Towards the automatic classification of traceability links. In: Automated Software Engineering (2017). https://doi.org/10.1109/ASE.2017.8115723

18. Pennington, J., Socher, R., Manning, C.D.: GloVe: global vectors for word representation. In: Empirical Methods in Natural Language Processing, pp. 1532–1543. ACL (2014). https://doi.org/10.3115/v1/d14-116
19. Robeer, M., Lucassen, G., van der Werf, J.M.E.M., Dalpiaz, F., Brinkkemper, S.: Automated extraction of conceptual models from user stories via NLP. In: Requirements Engineering, pp. 196–205. IEEE (2016). https://doi.org/10.1109/RE.2016.40
20. Salton, G.: Automatic Text Processing: The Transformation, Analysis, and Retrieval of Information by Computer. Addison-Wesley Longman Publishing Co., Inc., Boston (1989)
21. Schlutter, A., Vogelsang, A.: Knowledge representation of requirements documents using natural language processing. In: Natural Language Processing for Requirements Engineering. RWTH Aachen (2018). https://doi.org/10.14279/depositonce-7776
22. Schlutter, A., Vogelsang, A.: knowledge extraction from natural language requirements into a semantic relation graph. In: Knowledge Graph for Software Engineering). ACM (2020). https://doi.org/10.14279/depositonce-9772.2
23. Schlutter, A., Vogelsang, A.: Trace link recovery using semantic relation graphs and spreading activation. In: Requirements Engineering, pp. 20–31. IEEE (2020). https://doi.org/10.1109/RE48521.2020.00015
24. Shin, Y., Huffman Hayes, J., Cleland-Huang, J.: Guidelines for benchmarking automated software traceability techniques. In: Symposium on Software and Systems Traceability (2015). https://doi.org/10.1109/SST.2015.13
25. Singhal, A.: Modern information retrieval: a brief overview. Comput. Soc. Tech. Committee Data Eng. **24**(4), 35–43 (2001). http://singhal.info/ieee2001.pdf
26. Winkler, S., von Pilgrim, J.: A survey of traceability in requirements engineering and model-driven development. Softw. Syst. Model. **9**, 529–565 (2010). https://doi.org/10.1007/s10270-009-0145-0

CORG: A Component-Oriented Synthetic Textual Requirements Generator

Aya Zaki-Ismail[1]([envelope]) [ID], Mohamed Osama[1] [ID], Mohamed Abdelrazek[1] [ID],
John Grundy[2] [ID], and Amani Ibrahim[1] [ID]

[1] Information Technology Institute, Deakin University,
Burwood Highway, Burwood, VIC 3125, Australia
{amohamedzakiism,mdarweish,
mohamed.abdelrazek,amani.ibrahim}@deakin.edu.au
[2] Information Technology Institute, Monash University,
Wellington Road, Clayton, VIC 3800, Australia
John.Grundy@monash.edu

Abstract. The majority of requirements formalisation techniques operate on textual requirements as input. To establish and verify the reliability and coverage of such techniques, a large set of textual requirements with diverse structures and formats is required. However, such techniques are typically evaluated on only a few manually curated requirements that do not provide enough coverage of the targeted structures. Motivated by this problem, we introduce a Component-oriented synthetic textual requirements generator (CORG) that can generate large numbers of synthesised diverse-structure textual requirements, along with key components breakdowns. CORG utilises a controlled random-selection (CRS) strategy throughout the backtracking-based generation. We evaluate the coverage, diversity, performance and correctness of CORG. The evaluation results show that CORG can generate comprehensive diverse-structure combinations in reasonable time without being affected by the size of the produced requirements.

Keywords: Requirements engineering · Text generation

1 Introduction

Several requirements formalisation and extraction techniques [9,26,27] rely on textual requirements. The reliability of such techniques is critical, as they represent the foundation for the remaining requirements engineering and analysis tasks (e.g. 3C quality issues detection [26]). However, the evaluation of such techniques is typically limited to a curated set of few requirements [9,27]. The main drawbacks of this evaluation approach are:-

Zaki-Ismail and Osama are supported by Deakin PhD scholarships. Grundy is supported by ARC Laureate Fellowship FL190100035.

F. Dalpiaz and P. Spoletini (Eds.): REFSQ 2021, LNCS 12685, pp. 54–70, 2021.
https://doi.org/10.1007/978-3-030-73128-1_4

- Structure-biased requirements: In this case, few of the targeted structures are covered and essential ones required for an exhaustive evaluation may be missed [9]
- Small number of requirements: The number of requirements used in evaluation is, in many cases, relatively small. This is because real-world requirements are typically confidential and not published making them hard to obtain, and manually developed ones require a considerable amount of time and effort to develop.

An example of such limitations is present in the evaluation of the requirements formalisation approach described in [9]. This approach accepts textual requirements incorporating events, conditions and actions. However, the evaluation was performed on requirements with event-action and condition-action formats as indicated in Fig. 1: part *I*, in addition to some manually adjusted requirements from the (If A Then B) format to the (B If A) format as in Fig. 1: part *II*. However, requirements holding the popular event-condition-action behavioural requirements format [24] are completely missed (i.e., not used in any order) as shown in Fig. 1: part *III*.

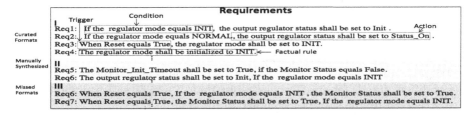

Fig. 1. Curated, synthesised and missed requirements samples by Gosh et al., [9]

In this paper, we tackle the problem of the utilisation of incomplete test data sets for the evaluation of formalisation techniques, by proposing a component-oriented synthetic textual requirements generator "CORG". It can be used to automatically generate comprehensive combinations of structurally-diverse synthesised textual requirements. In line with achieving such goal, all the generated requirements are expected to be equally useful for this problem without a need for a human analyst to assess their semantics. This is because formalisation techniques are insensitive to the semantics of the input requirements (i.e., their underlying analysis depends mainly on the requirements' grammatical syntax to be transformed into a corresponding formal notation [9,26]). In addition, what enables a reliable evaluation of such techniques, is the co-existence of diverse structures within a combinatorially complete set of requirements.

We also propose the output layout of the synthesised requirements and the corresponding formal grammar. CORG adheres to this grammar during the generation process to produce requirements in the target output layout. The grammar and layout are based on a Requirement Capturing Model (RCM) [28]. RCM is a reference model that incorporates the key properties (i.e., the key components and sub-components) that may exist in a system requirement sentence. It extends popular requirements expression formats (e.g., EARS[17], etc.).

We evaluated the capabilities of CORG in generating: (1) a complete set of the possible combinations of requirement's properties, (2) diverse structures, (3) correct requirements, and (4) realistic requirements. In addition to evaluating the generation time.

2 Background

RCM [28] is a semi-formal representation model that aggregates the behavioural NL-requirements components defined in literature and provides their respective sub-components and arguments breakdown. It supports a wide range of requirements because the model adapts to any permutation of its components.

RCM supports four requirements component types (i.e., Fig. 1 shows samples):

- Action: expresses the tasks performed by the system.
- Trigger: represents events that implicitly fire actions within the system cycle.
- Condition: stands for specific constraints that should be satisfied and explicitly checked by the system for an action to occur.
- Conditional scope: represents the governing conditions required for either checking the preconditions (triggers and conditions) called "preconditional scope" or performing the action(s) "called action scope".

Sub-component types associated with the above components:

- valid-time: the period of time for a component to be valid (e.g., the inhibitor shall transition to [True] <u>for at most 1 s</u>).
- pre-elapsed-time: the consumed time –from a reference point– before a component starts (e.g., <u>within 2 s</u> the IDC transitions to [True]).
- in-between-time: the time between two consecutive repetitions of an event (e.g., the signal turns to [true] <u>every 2 s</u>).
- hidden-constraint: a constraint held for only one argument (i.e., system entity) within the component (e.g., **the entry** <u>whose index exceeds 2</u> shall be incremented, the underlined text is held for the bold text).

Table 1 shows the possible sub-components of each component type.

Table 1. Sub-components association with each component type

Sub-components	Components				
	PreConditional-Scope	Action-Scope	Condition	Trigger	Action
Hidden Constraint	✓	✓	✓	✓	✓
Valid-time	✓	✓	✓	✓	✓
Pre-elapsed-time			✓		✓
In-between-time				✓	✓

3   CORG Formal Grammar

We designed the output layout (targeting RCM) for the generated requirements, components, sub-components, and arguments breakdowns as follows:-

Listing 1.1. Generated Requirements Output Layout

```
req(reqText, CompList).
CompList::= [comp(SubCompList, compText),...]
SubCompList::= [subComp(Type, BreakDowns, subCompText), ...]
Type::= trig|cond|precondScope|actScope|act|hidden|v-time|pre-time|in-time
BreakDowns::=  TimeInfo|RelClause|SubClause|Clause
TimeInfo::= [prePosition, quanitfication, value, unit, Nil]
RelClause::= [relNoun, relPronoun, subj, verb, [complement1, complement2,...] ]
SubClause::= [Nil, head, subj, verb, [complement1,complement2,...]]
Clause::= [Nil, Nil, subj, verb, [complement1,complement2,...]]
```

"name()" indicates composite entity, "[]" is list representation, and "Nil" means an empty item in the list. "req()" is a composite entity representing a requirement sentence, "reqText" is the text of the generated sentence, and "CompList" is a list of component entities each represented in "comp()". "comp-Text" holds component text, and "SubCompList" is a list of sub-components entities each represented in "subComp()". "Type" is the (sub)component type, "subCompText" is the sub-Component text, and "BreakDowns" is a TimeInfo, RelClause, SubClause, or Clause according to the sub-component's type (all of them are lists of 5 items storing the sub-component arguments, where each item has a specific role based on its position in the list). "TimeInfo" represents breakdowns of time-related RCM subcomponents (i.e., valid-time, pre-elapsed-time and in-between-time). "RelClause", "SubClause", and "Clause" represent the breakdowns of the RCM hidden constraint sub-component, condition/trigger/condional-Scope, and action component respectively.

Then, we developed a formal grammar to govern the generation process in line with the output layout. We only support present, future, imperative tenses and active/passive voices with correct syntax according to the English grammar. We define the formal grammar of the supported structures as follows:-

Listing 1.2. CORG Formal Grammar

```
<Sentence> ::= <Subclause>*.<Clause>.<Subclause>*
<Subclause> ::= Subordinator.<clause>
<Clause> ::= [<Subject>].[<RelClause>].<Predicate>.<TimeInfo>*
<RelClause> ::= HiddenConstHead.[Property].<Predicate>
<Subject> ::= <NounPh>
<NounPh> ::= Noun.<Modifier>*
<Modifier> ::= Preposition.<NounPh>
<Predicate> ::= [Modality].<MainVerb>.<Complement>+
<MainVerb> ::= Verb |(be).Verb.(ed)
<Complement> ::= [Preposition].(Noun|SystemValue)
Modality ::= "shall"|"will"|...
Subordinator ::= ConditionHead|TriggerHead|ScopeHead
ConditionHead ::= "if"|"provided that"|...
TriggerHead ::= "when"|"once"|"whenever"|...
ScopeHead ::= "after"|"before"|"until"|"while"|...
HiddenConstHead ::= "whose"|"that"
<TimeInfo> ::= TimePreposition. [QuantifyingRel]. Value. Unit
TimePreposition ::= Valid-Time-Prep|Pre-elapsed-Time-Prep|In-Time-Prep
Valid-Time-Prep ::= "for"|"up to"|...
Pre-elapsed-Time-Prep ::= "within"|"in"|"after"|...
In-Between-Time-Prep ::= "every"|...
QuantifyingRel ::= "less than"|"less than or equal"|"at most"|...
Value ::= Number
Unit ::= "seconds"|"minuets"|"milliseconds"|...
```

where, "*" indicates the presence of zero or more items, "+" means one or more, "." means the composition of different items, "..." means other words in the input dictionary, "< >" means non-terminal, and "[ ]" means optional item. Nouns, Verbs, Properties, SystemValues and Prepositions are not further decomposed (terminals) and are fetched from the input dictionary. The proposed grammar allows temporal operators through the element subordinator. A requirement sentence consists of at least one clause. A clause consists of at least a main predicate expressing the core meaning of the sentence, and optionally a subordinator –conditional or temporal conditional head– can be attached to extend the meaning as a subordinating clause.

4   CORG

CORG takes as input: (1) a dictionary of the domain lexical words and verb frames, and (2) size of requirements to be generated. It then utilises backtracking (a well-known approach previously adopted in textual generation [20]) through the built-in backward chaining inference engine of the Prolog programming language (a descriptive logic programming language consisting of a set of definite clauses (facts and rules) correlated to artificial intelligence and computational linguistics [22]).

Backtracking is typically a depth first search (DFS) mechanism, in which, an arbitrary decision is made at each choice-point. When a dead-end is reached, the inference engine backtracks to the last decision-point that can have a different path, makes a different choice, then proceeds from there. It can iterate over all the possible arrangements of a search space and provide all combinations and permutations [11]. However, it exploits all the possible permutations of the combination at hand before transitioning to the next combination. CORG performs a controlled random selection CRS (i.e., based on Prolog equi-distributed randoms [16]) at specific choice-points. This mitigates the exploitation pattern of DFS, and ensures combinations comprehensiveness and diversity maintenance even at small generation sizes while preventing structurally broken combination(s) (e.g., a requirement without an action is in-correct).

For each element in the (sub-)components set, CRS assigns a 0 or 1 random number to decide its inclusion/exclusion in the current combination. The combination is then rearranged to allow permutations and maintain generation diversity. The underlying equi-distribution allows CRS to produce a different combination at each call (i.e., ensures the coverage of the possible combinations whenever the generation size exceeds the combinations count). The generation then goes through Java APIs, SimpleNLG [8], and Stanford-NLP [15] as in Fig. 2.

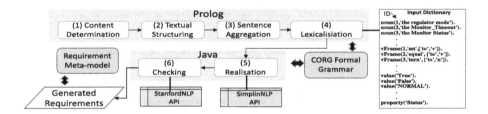

Fig. 2. CORG Framework

Text generation typically consists of five main tasks: content determination, textual structuring, sentence aggregation, lexicalisation, realization[7]. These tasks are applied differently according to the nature of the addressed problem. The underlying text generation approach adopted by CORG has been widely used in addressing a similar problem (i.e. sentence generation for quality testing and reliability evaluation) where several attempts have been carried out to generate textual sentences to test programming languages compilers [14,23,29], and regular expressions [25,30]. We also add a checking task to ensure fault free generation. The first four tasks are implemented with Prolog and the remaining two are implemented with Java as in Fig. 2. CORG follows the proposed formal grammar and generates requirements in the proposed format. The six tasks are described in the following subsections and supported with a step by step generation example in Fig. 3.

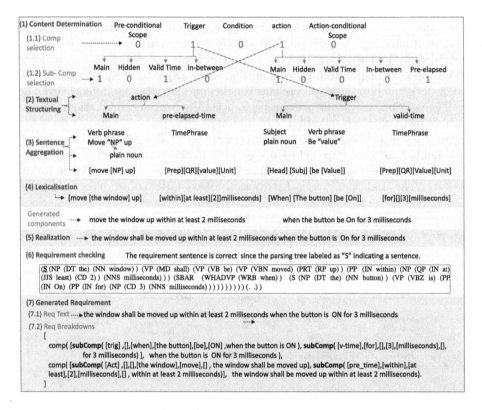

Fig. 3. Step by step CORG generation example.

4.1 Content Determination

This task determines which components and sub-components will be included in the generated requirement R. It consists of two levels, the selection(s) made at each level lead(s) to different choices in the next level(s). At the **first level**, CRS is applied on the components set: Preconditional scope, condition, trigger, action scope except the mandatory action component as in Fig. 3.1.1. In the **second level**, CRS is applied on the sub-comoponents set: different for each chosen component in the previous level as in Fig. 3.1.2 (i.e., Table 1 for sub-components association). Eventually, the count and types of components and sub-components contributing in the generated requirement are identified. This task embodies the first four lines of the formal grammar and the CompList, and SubCompList in the output layout of the generated requirements.

Components and sub-components available for selection can be controlled before the generation process through CORG settings. This allows for adaptation into different domains and usage scenarios (*e.g.*, CIRCE [1] uses only event-condition-action ECA components).

4.2 Textual Structuring

This task determines the order of the components within the sentence and the sub-components within the component. Alternative arrangement/permutation can be achieved through the random reordering technique. The approach takes one element (of a given combination) at a time and inserts it at a random free position in the new version as indicated in Algorithm 1.3. Similar to CRS, the random reordering is capable of providing a different permutation at each call (i.e., which ensures the comprehensiveness of the permutations for the same combination in large enough sizes and maintains diversity in small sizes). Figure 3.2 shows the reordered components after applying random reordering.

Listing 1.3. Random Reordering Algorithm

```
OutComps = φ
While(InComps ≠ φ){
    CrrComp = InComps.removeFirst()
    Len = OutComp.length()
    RandomIdx = getRandom(0, Len+1)
    OutComp.insertAt(RandomIdx, CrrComp)
}
```

4.3 Sentence Aggregation

This task selects the grammatical structures to apply on an individual component. By the end of this task, the outlined formal grammar clauses in Task1 shall be assigned a complete random grammatical structure as in Fig. 3.3. A clause may include up to four different types of parts (i.e., Subject, RelClause, Predicate and TimeInfo as in Grammar 1.2). Each part has more than one valid structure. Table 2 lists the alternative structures of each part conforming to the proposed formal grammar, where NounSize and VFrameSize indicate the count of distinct nouns and verb-frames in the dictionary respectively. In addition, it highlights how these structures are used to select a new random structure.

Table 2. Alternative structure for clause's breakdowns Controlled in CORG

Type	Grammatical Rules Pseudo	Type	Grammatical Rules Pseudo

Subject	if nominalCount = 1 then
	    ID ← getRandom(1,NounSize)
	    Noun ← getNominal(ID)
	    return Noun
	else
	    Prep ← getCompositionPrep()
	    Noun ← Prep + getNominal(nominalCount -1)
	    return Noun
	end if
		Predicate	Dictionary selection: the grammatical frames of each verb stored in the dictionary
			    ID ← getRandom(1,VFrameSize)
			    VFrame ← selectFrame(ID)
			    return VFrame

TimeInfo	if Type = ValidTime then
	    Prep ← getRandVTimePrep()
	else if Type = PreElapsed then
	    Prep ← getRandPreTimePrep()
	else if Type = InTime then
	    Prep ← getRandInTimePrep()
	end if
	QR ← getRandQantRel()
	Val ← getRandValue()
	Unit ← getRandUnit()
	TI ← aggregat(Prep,QR,Val,Unit)
	return TI
		RelClause	if Type = OnProperty then
			    RelP ← getPropRelPronoun()
			    Prop ← getRandProperty()
			else if Type = OnNoun then
			    RelP ← getNounRelPronoun()
			    Property ← φ
			end if
			VFrame ← getVerbFrame()
			RVF ← realizeFrame(VFrame)
			RC ← aggregat(RelP,Prop,RVF)
			return RC

4.4   Lexicalisation

In this task, the chosen grammatical rules in the previous task are populated with randomly selected lexical words conforming to the grammatical roles as in Fig. 3.4. These words are fetched from the dictionary. Table 3 lists the structures (each has a grammatical role) in the dictionary with descriptions. Each structure instance has an "Id" to allow random selection from the same structure type (i.e., a random number is generated to fetch the lexical word whose id equals the generated number and whose structure (grammatical role) is regulated by the syntactic rules as indicated in Table 2). By the end of this task, the sub-components' breakdowns in the output layout shall be complete.

Table 3. Dictionary Metamodel

Structure	Description	Example
noun(*ID, Noun*)	Nominal Noun	noun(1, 'the car')
vFrame(*ID, Verb, ArgList*)	verb-frame representing the predicate structure. -*ArrgList: expresses the verb associated: arguments notations and prepositions in order within the frame *Arg-notations: ('v' → sys-value) and ('n' → noun)	vFrame(1, 'set', ['to', 'v'])
property(*ID, Property*)	Property for nominal nouns used in relative clauses	property(1, 'door')
sysValue(*ID, Value*)	domain values for nominal nouns or properties	sysValue(1, '[Locked]')

4.5   Realisation

The generated components for each requirement require tense adjustment as per the English grammar. This task considers adjusting the tense of each component (each expressed by a clause). In this task, all components' types are assigned to the present tense except actions (future tense), as shown in Algorithm 1.4. For diversity, we put the action in the imperative form using active and passive voices randomly. Figure 3.5 shows an example of the components after tense adjustment. We adjust the tense using SimpleNLG. *First*, we prepare the subject, verb and complement of the current component. *Then* we feed them to the SimpleNLG realiser along side other grammatical flags (*e.g.*, tense(in both levels), voice, person). *Finally*, we readjust the affected parts in the requirement and their breakdowns.

Listing 1.4. Tense Adjustment

```
If(Comp.Type = "act")
    Voice = get random voice
    If(Voice = "Active"}
        adjustToImperative(Comp)
    Else
        adjustToPassiveFuture(Comp)
Else
    adjustToActivePresent(Comp)
```

4.6   Requirements Checking

In the generated requirement a "," is appended to all the components (except the last one). Depending on the generated components, random re-ordering may cause grammatical errors because of the incorrect punctuation. We use Stan-fordNLP to detect ungrammatical sentences by analysing the output of the parse tree. The parse tree represents the syntactic structure of the given string according to a specific context-free-grammar. A requirement sentence whose obtained parsing tree is labeled with an "s" (i.e., sentence) is correct [21]. Figure 4 shows two parsing tree (a) and (b) of the same requirement (with different punctuation), tree in (b) is grammatically correct while the other one in (a) is incorrect. (a) can be corrected by removing incorrect comma(s) as in (b). In general it is not recommended to depend on the output of the parse tree alone in such checking. However, in our case it is effective because the realisation task ensures the correctness of each component. According to Fig. 3.6, the generated requirement in the tracing example is correct.

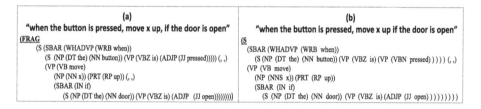

Fig. 4. Parsing trees of (the same requirement sentence with different punctuation)

5   Evaluation

In this section, we evaluate CORG[1] using assessment criteria (coverage and time performance metrics) that have proven effective in evaluating text generation in several approaches [14,25,30]. We conducted five experiments to evaluate CORG on such metrics in addition to assessing diversity, correctness, and realisticness:

- Coverage: Does CORG provide comprehensive combinations of the components and sub-components according to RCM?
- Time Performance: How does the requirements size affect CORG's generation time?
- Diversity: Is CORG capable of providing all the possible arrangements of each combination?
- Correctness: Does CORG correctly generate requirements as expected?
- Realisticness: Can CORG generate semantically sound requirements

The dictionary[2] used in the generation contains 16 nominal nouns, 7 verb frames, 4 values and one property (i.e., obtained from requirements used in [9]).

---

[1] CORG Source Code: https://github.com/ABC-7/CORG/tree/main/CORG.

[2] Dictionary and Generated-requirements:https://github.com/ABC-7/CORG/tree/main/Experiment.

5.1   Generation Coverage

In this experiment, we generated 500 unique requirements (see Footnote 2) since the possible combinations of components and sub-components are 448 (i.e., calculated using ^nC_r). The basis of our evaluation is the correct combinations of the possible power set of components and sub-components that can constitute one requirement. We evaluate the coverage on two levels.

First level tests the coverage of the possible correct components combinations. Figure 5(a) shows a Venn-diagram highlighting in dark the correct combinations of components among all the possible combinations (i.e., a requirement without action is incorrect). Figure 5(b) shows the percentage of each components combination within the generated requirements. It can be seen that, the generated requirements cover all the correct combinations and do not incorporate any incorrect combinations (e.g., a requirement with no action).

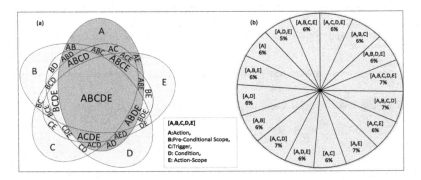

Fig. 5. (a) Venn-diagram for components combination (correct combinations highlighted in dark gray). (b) Components combinations percentages in generated requirements

In the second level, we evaluate the coverage of the sub-components of each individual component. Figure 6 shows the ability of CORG to cover the generation of the possible sub-components combinations (i.e., where, the action has $2^4 = 16$ possible combinations of its associated sub-components, condition and trigger have $2^3 = 8$, and the other components have $2^2 = 4$ combinations). Each column shows the percentages corresponding to each combination of the sub-components for the intended component within the entire requirements. It can be seen that, no combination is missed.

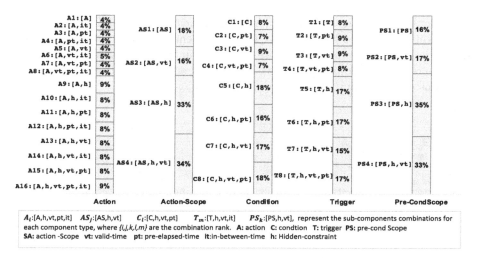

A1:[A]	4%			C1:[C]	8%	T1:[T]	8%		
A2:[A,it]	4%	AS1:[AS]	18%	C2:[C,pt]	7%	T2:[T,pt]	9%	PS1:[PS]	16%
A3:[A,pt]	4%								
A4:[A,pt,it]	4%			C3:[C,vt]	9%	T3:[T,vt]	9%		
A5:[A,vt]	4%								
A6:[A,vt,it]	5%	AS2:[AS,vt]	16%	C4:[C,vt,pt]	7%	T4:[T,vt,pt]	8%	PS2:[PS,vt]	17%
A7:[A,vt,pt]	4%								
A8:[A,vt,pt,it]	4%								
A9:[A,h]	9%			C5:[C,h]	18%	T5:[T,h]	17%		
A10:[A,h,it]	8%	AS3:[AS,h]	33%					PS3:[PS,h]	35%
A11:[A,h,pt]	8%			C6:[C,h,pt]	16%	T6:[T,h,pt]	17%		
A12:[A,h,pt,it]	8%								
A13:[A,h,vt]	8%			C7:[C,h,vt]	17%	T7:[T,h,vt]	15%		
A14:[A,h,vt,it]	8%	AS4:[AS,h,vt]	34%					PS4:[PS,h,vt]	33%
A15:[A,h,vt,pt]	8%			C8:[C,h,vt,pt]	18%	T8:[T,h,vt,pt]	17%		
A16:[A,h,vt,pt,it]	9%								
Action		**Action-Scope**		**Condition**		**Trigger**		**Pre-CondScope**	

A_i:[A,h,vt,pt,it] AS_j:[AS,h,vt] C_l:[C,h,vt,pt] T_m:[T,h,vt,it] PS_k:[PS,h,vt], represent the sub-components combinations for each component type, where {i,j,k,l,m} are the combination rank. **A**: action **C**: condition **T**: trigger **PS**: pre-cond Scope **SA**: action -Scope **vt**: valid-time **pt**: pre-elapsed-time **it**:in-between-time **h**: Hidden-constraint

Fig. 6. The possible sub-components combinations coverage

5.2   Time Performance

We evaluated CORG generation time (on Prolog) for unique and redundant requirements on ten different data-set sizes. We measured the average time of five samples for each data-set size in both (unique and redundant data-sets). Figure 7 shows that CORG generates up-to 10000 unique requirements in around seven seconds and around one second for the redundant

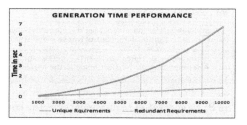

Fig. 7. CORG avg. Generation time

requirements. Unique data-set generation requires more time since more time is required to eliminate and replace redundant generation attempts with unique ones. The unique generation time depends on the dictionary size (larger dictionary sizes guarantee less generation time especially for larger data-set sizes).

5.3   Diversity Evaluation

The textual structuring step in Sect. 4.2 is responsible for maintaining structure diversity within the generated requirements (i.e., by getting different permutation each time for the given combination). To assess diversity, we set an experiment to generate all requirements holding one combination, then assess diversity (i.e., permutations) within the generated requirement. The presumed combination is a requirement with condition, trigger and action components (i.e., any generated requirement would have the three components). This combination has six permutations: {(Cond,Act,Trig), (Cond,Trig,Act),

(Trig,Act,Cond), (Trig,Cond,Act), (Act,Trig,Cond), (Act,Cond,Trig)}. Finally, we informed CORG to generate just 10 requirements with this setting. Figure 8 shows that, the six arrangements are generated in the first seven requirements (i.e., the remaining requirements have repeated arrangements). It is worth noting that, this displayed output is before the tense adjustment step.

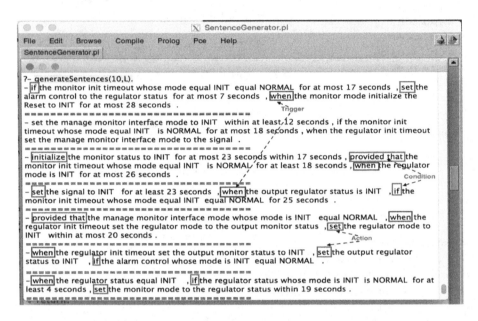

Fig. 8. 10 generated requirements for the combination (condition, trigger, action) highlighting the corresponding complete six permutations

5.4   Correctness Evaluation

We feed the 500 requirements generated in the coverage experiment to an automated NLP-based requirements extraction approach [28]. Then, we automatically compared the generated breakdowns (i.e., components, sub-components and, arguments) to the extracted ones[3] through string matching. The experiment shows that all requirements are generated correctly as expected. The used extraction approach depends mainly on StanfordNLP typed dependency which has a percentage of error [28]. To provide the extraction with fully correct interpretations, the sub-components text of each generated sentence are addressed apart by StanfordNLP and their corresponding typed dependencies are aggregated. Finally, the aggregated typed dependencies are used in the extraction.

---

[3] Extraction-Output: https://github.com/ABC-7/CORG/blob/main/Extractionlog. xml.

5.5   Realisticness Evaluation

We evaluated the ability of CORG to generate realistic requirements (similar to human-written ones). To achieve this, we fed the tool with the dictionary of a group of manually-specified requirements for a target system. Then, we used CORG to check if these requirements can be generated or not - reverse engineering utilising the inference engine of Prolog to check if a given output could be produced from the given input.

We conducted the experiment on a data-set of 19 requirements collected from the literature - used in [9]. We fed CORG with (a) the requirements dataset and (b) the system dictionary. CORG successfully constructed the breakdown of all input requirements[4]. This experiment shows that a subset of the generated requirements is both syntactically and semantically correct.

5.6   Strengths and Limitations

The key strengths of CORG are: (1) providing combinatorially complete coverage for components and sub-components, (2) allowing the customisation of components and sub-components contributing in the generation process, (3) ensuring structure-diversity in small and large data-set sizes, (4) associating the generated requirements with their breakdowns and (5) large number of requirements may be generated from a small dictionary with very few details (i.e., just lexical words and verb frames). It is also worth noting that CORG can be easily enhanced to ensure the generation of semantically reasonable requirements by adding association rules to the input dictionary to only allow certain lexical words and verb frames to be selected together.

The main limitation of CORG is that the generated requirements are not all semantically reasonable because the concrete system relations are not considered in the dictionary. However, this does not affect the effectiveness of the generated requirements in enabling a reliable evaluation of the formalisation approaches because such approaches are only affected by the syntax of the input.

6   Related Work

We cover the related work from three perspectives:

(a) **sentence generation from lexical words:** data driven textual generation is a widely used text generation approach (HALogen [13], Nitrogen [12] and FERGUS [2]. These approaches adopt a two-stage architecture. In the first stage, a forest of the possible expressions is constructed. In the second stage, the expressions are selected using a probabilistic model. In contrast, CORG applies CRS instead of the forest construction to save generation time. GENERATE [10] randomly generates sentences using a small set of English phrases, syntactic

---

[4] Input-requirements, Used-dictionary, and Resulting-breakdowns for realisticness experiment: https://github.com/ABC-7/CORG/tree/main/ValidationExperiement.

rules and transformation rules to form valid sentences. As input, it takes a dictionary containing both the nouns and verbs (semantically coded to assure that invalid sentences such as "The building smoked a cigar" will not be produced). The dictionary consists of twenty verbs and twenty nouns. For flexibility, CORG supports dictionaries with or without semantics.

(b) Requirements generation: most approaches generate textual requirements from software engineering models [19]. In [18], NL requirements are generated from UML class diagrams. This approach uses a rule set in conjunction with a linguistic ontology to express the components of the diagram. Alternatively, the approach presented in [4] relied on system domain-specific grammar to provide description and information regarding specific requirements technical terms. In [5] the Semantics of Business Vocabulary and Business Rules (SBVR) was used as an intermediate representation for transforming UML into constrained natural language. Similarly, CORG uses a defined grammar for the generated requirements controlled by a set of rules. However, the content of the generated requirement(s) is syntactically correct but not limited by a specific system since no relations are enforced in the dictionary. Other approaches generate creative requirements (i.e., more useful and novel requirements) for the sustainability of software systems. Bhowmik et al [3] propose a framework to obtain creative requirements by making unfamiliar connections between familiar possibilities of requirements. In [6], a novel framework generates creative requirements utilising NLP and ML techniques for both novel and existing software. The framework reuses requirements from similar software in the application domain and leverages the concept of requirement boilerplate to generate candidate creative requirements. Such approaches include a manual checking process to discard useless outcome. Similarly, CORG eliminates useless requirements –syntax oriented– through an automatic checking process.

(c) Text generation for evaluation: several attempts [14, 23, 29] have been carried out to generate sentences to test the parsing of programming languages compilers. In addition, textual strings are generated in [25, 30] to evaluate regular expressions. The main feature of these techniques is that the generated text must obey to the formal grammar of the compiler/regular expression. Similarly, requirements generated by CORG follow a formal grammar for describing systems behavior. We share with such approaches the goals and metrics of the generated text (i.e., combinations coverage for robust evaluation and generation performance).

7   Conclusion

In this paper, we introduced CORG; a synthetic requirements generator that can produce all the possible combinations and diverse structures with respect to the RCM set of key requirements properties. First, we defined a formal grammar for the generated requirements. Then, we employed the backtracking technique with a controlled random-selection to ensure combinatorial comprehensiveness and maintain diversity in small data-sets. Evaluation results show that CORG is able to generate comprehensive combinations with diverse structures regardless of the size of the generated requirements. In the future, we aim to investigate CORG capabilities in: (1) generating requirements in other languages, (2) filtering semantically unreasonable requirements utilising both human vetting and dictionary rules (which can be useful for generating creative requirements).

References

1. Ambriola, V., Gervasi, V.: On the systematic analysis of natural language requirements with circe. Autom. Softw. Eng. **13**(1), 107–167 (2006)
2. Bangalore, S., Rambow, O.: Exploiting a probabilistic hierarchical model for generation. In: Proceedings of the 18th Conference on Computational Linguistics, vol. 1, pp. 42–48. Association for Computational Linguistics (2000)
3. Bhowmik, T., Niu, N., Mahmoud, A., Savolainen, J.: Automated support for combinational creativity in requirements engineering. In: 2014 IEEE 22nd International Requirements Engineering Conference (RE), pp. 243–252. IEEE (2014)
4. Burden, H., Heldal, R.: Natural language generation from class diagrams. In: Proceedings of the 8th International Workshop on Model-Driven Engineering, Verification and Validation, pp. 1–8 (2011)
5. Cabot, J., Pau, R., Raventós, R.: From UML/OCL to SBVR specifications: a challenging transformation. Inf. Syst. **35**(4), 417–440 (2010)
6. Do, Q.A., Bhowmik, T., Bradshaw, G.L.: Capturing creative requirements via requirements reuse: a machine learning-based approach. J. Syst. Softw. **170**, 110730 (2020)
7. Gatt, A., Krahmer, E.: Survey of the state of the art in natural language generation: core tasks, applications and evaluation. J. Artif. Intell. Res. **61**, 65–170 (2018)
8. Gatt, A., Reiter, E.: SimpleNLG: a realisation engine for practical applications. In: Proceedings of the 12th European Workshop on Natural Language Generation (ENLG 2009), pp. 90–93 (2009)
9. Ghosh, S., Shankar, N., Lincoln, P., Elenius, D., Li, W., Steiener, W.: Automatic requirements specification extraction from natural language (arsenal). Technical report, SRI INTERNATIONAL MENLO PARK CA (2014)
10. Hackenberg, R.G.: Generate: a natural language sentence generator. CALICO J. **2**(2), 5–8 (2013)
11. Khachiyan, L., Boros, E., Elbassioni, K., Gurvich, V.: A new algorithm for the hypergraph transversal problem. In: Wang, L. (ed.) COCOON 2005. LNCS, vol. 3595, pp. 767–776. Springer, Heidelberg (2005). https://doi.org/10.1007/11533719_78
12. Langkilde, I.: Forest-based statistical sentence generation. In: Proceedings of the 1st North American chapter of the Association for Computational Linguistics Conference, pp. 170–177. Association for Computational Linguistics (2000)
13. Langkilde-Geary, I.: An empirical verification of coverage and correctness for a general-purpose sentence generator. In: Proceedings of the International Natural Language Generation Conference, pp. 17–24 (2002)
14. Lutovac, M.M., Bojić, D.: Techniques for automated testing of Lola industrial robot language parser. Telfor J. **6**(1), 69–74 (2014)
15. Manning, C.D., Surdeanu, M., Bauer, J., Finkel, J.R., Bethard, S., McClosky, D.: The Stanford CorenLP natural language processing toolkit. In: Proceedings of 52nd Annual Meeting of the Association for Computational Linguistics: System Demonstrations, pp. 55–60 (2014)

16. Matsumoto, M., Nishimura, T.: Mersenne twister: a 623-dimensionally equidistributed uniform pseudo-random number generator. ACM Trans. Model. Comput. Simul. (TOMACS) **8**(1), 3–30 (1998)
17. Mavin, A., Wilkinson, P., Harwood, A., Novak, M.: Easy approach to requirements syntax (ears). In: 17th IEEE International Requirements Engineering Conference, RE 2009, pp. 317–322. IEEE, August 2009
18. Meziane, F., Athanasakis, N., Ananiadou, S.: Generating natural language specifications from UML class diagrams. Requirements Eng. **13**(1), 1–18 (2008)
19. Nicolás, J., Toval, A.: On the generation of requirements specifications from software engineering models: a systematic literature review. Inf. Softw. Technol. **51**(9), 1291–1307 (2009)
20. O'Donnell, M.: Sentence analysis and generation-a systemic perspective. Ph.D. thesis, University of Sydney (1994)
21. Osama, M., Zaki-Ismail, A., Abdelrazek, M., Grundy, J., Ibrahim, A.: Score-based automatic detection and resolution of syntactic ambiguity in natural language requirements. In: 2020 IEEE International Conference on Software Maintenance and Evolution (ICSME), pp. 651–661. IEEE (2020)
22. Pereira, F.C., Shieber, S.M.: Prolog and Natural-Language Analysis. Microtome Publishing (2002)
23. Purdom, P.: A sentence generator for testing parsers. BIT Numer. Math. **12**(3), 366–375 (1972)
24. Qiao, Y., Zhong, K., Wang, H., Li, X.: Developing event-condition-action rules in real-time active database. In: Proceedings of the 2007 ACM Symposium on Applied Computing, pp. 511–516. ACM (2007)
25. Radanne, G., Thiemann, P.: Regenerate: a language generator for extended regular expressions. In: Proceedings of the 17th ACM SIGPLAN International Conference on Generative Programming: Concepts and Experiences, pp. 202–214. ACM (2018)
26. Yan, R., Cheng, C.H., Chai, Y.: Formal consistency checking over specifications in natural languages. In: 2015 Design, Automation & Test in Europe Conference & Exhibition (DATE), pp. 1677–1682. IEEE (2015)
27. Zaki-Ismail, A., Osama, M., Abdelrazek, M., Grundy, J., Ibrahim, A.: RCM-extractor: automated extraction of a semi formal representation model from natural language requirements. In: Proceedings of the 9th International Conference on Model-Driven Engineering and Software Development (2021)
28. Zaki-Ismail, A., Osama, M., Abdelrazek, M., Grundy, J., Ibrahim, A.: RCM: requirement capturing model for automated requirements formalisation. In: Proceedings of the 9th International Conference on Model-Driven Engineering and Software Development (2021)
29. Zelenov, S.V., Zelenova, S.A.: Generation of positive and negative tests for parsers. Program. Comput. Softw. **31**(6), 310–320 (2005)
30. Zheng, L., Ma, S., Wang, Y., Lin, G.: String generation for testing regular expressions. Comput. J. **63**, 41–65 (2019)

Automatically Classifying Non-functional Requirements with Feature Extraction and Supervised Machine Learning Techniques: A Research Preview

Mahtab EzzatiKarami[✉] and Nazim H. Madhavji

The University of Western Ontario, London, ON N65B7, Canada
mezzati@uwo.ca

Abstract. Context and Motivation: In large projects, extracting the relevant NFR-information as per the stakeholder's responsibility and needs can be time-consuming and challenging. **Question/Problem:** Classification of NFRs is one way to mitigate this problem. However, because of the size and complexity of the SRS, the manual classification of NFRs is considered time-consuming, labour-intensive, and error-prone. An automated solution is needed that provides a reliable and efficient classification of NFRs. **Principal ideas/results:** Using natural language processing and supervised machine learning (SML) algorithms, we investigate feature extraction techniques (i.e., POS-tagging based, BoW, and TF-IDF) to assess their efficacy in automated classification, in conjunction with the SML algorithms (such as: SVM, SGD SVM, LR, DT, Bagging DT, Extra Tree, RF, GNB, MNB, and BNB). **Contribution:** The proposed combinations: (i) SVM with TF-IDF, (ii) LR with POS and BoW, and (iii) MNB with BoW, all achieve precision and recall values greater than 0.85, and process execution time of less than 0.1 s. Comparison with related work is favourable as is preliminary validation using an industry dataset.

Keywords: Non-functional requirements · Classification · Supervised Machine Learning · Feature extraction

1 Introduction

Non-functional requirements (NFRs) describe desirable quality attributes (e.g., performance, reliability, and availability) of a software system. In a software requirements specification (SRS) document, the functional requirements and NFRs are often mixed together (perhaps categorised under domain or application-specific headers). In large projects, understanding and extracting the relevant NFR-information as per the stakeholder's responsibility and needs can be time-consuming and challenging, due to the size, complexity, and lack of familiarity with the SRS.

For example, in one large project [1], there were approx. 600 regulatory requirements buried amongst 12,000 requirements spread over a thousand pages of a contract. In such

© Springer Nature Switzerland AG 2021
F. Dalpiaz and P. Spoletini (Eds.): REFSQ 2021, LNCS 12685, pp. 71–78, 2021.
https://doi.org/10.1007/978-3-030-73128-1_5

cases, manually identifying the needed NFRs for project-work is effortful and error-prone [1]. Other researchers [2] have also shown that classifying NFRs into different types can aid stakeholders' project concerns [2].

In our research, we investigated how accurately we can classify NFRs automatically into various types; in particular, *usability*, *security*, *performance*, and *operational* requirements – which are the top four NFR-types in the PROMISE dataset. The goal of our research is to provide stakeholders with a reliable and efficient solution for extracting specific NFRs from SRS documents using natural language processing techniques combined with supervised machine learning (SML) algorithms. We investigate feature extraction techniques (i.e., POS-tagging based, BoW, and TF-IDF) to assess their efficacy in automated classification, in conjunction with the SML algorithms (such as: SVM, SGD SVM, LR, DT, Bagging DT, Extra Tree, RF, GNB, MNB, and BNB).

The resultant combinations: (i) SVM with TF-IDF, (ii) LR with POS and BoW, and (iii) MNB with BoW, all achieve precision and recall values greater than 0.85 and process execution times of less than 0.1 s; meaning, the right NFRs would be rendered, and quickly, to the stakeholder concerned. Comparison with related work [2, 4] is favourable as is the classification of NFRs in an industry dataset.

2   Related Work

We describe and analyse several previous attempts at classifying NFRs using machine learning techniques. In 2006, Cleland-Huang et al. [2] classified 370 NFRs from 15 SRS documents developed by graduate students. Their used certain keywords as indicator terms to distinguish different types of NFRs, trained these, and then classified NFRs from given documents according to the occurrence of the indicator terms. Their model yielded a classification recall of approx. 0.8 (considered high) and precision of 0.12 (low) due to a high rate of false positives.

In 2011, Zhang et al. [3] conducted an empirical study on classifying NFRs using SVM and the PROMISE dataset. Three kinds of index terms, at different levels of linguistic semantics (N-grams, individual words, and multi-word expressions (MWE)) were used in the representation of NFRs. Their results show that index terms as individual words with Boolean weighting outperform N-grams and MWEs; and using MWEs does not enhance the representation of individual words significantly. Also, they observed that automatic classification results in better performance on categories of large sizes than of small sizes. They conclude that individual words are the best index terms in text representation of short NFR descriptions. In comparison to the study by Cleland-Huang et al. [2], they report higher precision but lower recall values.

In 2017, Lu et al. [4] classified app user reviews into four types of NFRs (i.e., reliability, usability, portability, and performance), functional requirements, and others. Their approach combines four classification techniques (BOW, TF-IDF, CHI2 and AUR-BOW) with three machine learning algorithms (Naïve Bayes (NB), J48, and Bagging) to classify the reviews. The combination of AUR-BOW with Bagging achieved the best result (precision: 0.71 and recall: 0.72). An interesting observation made was that automatic classification using an imbalanced dataset performs poorly when the numeracy of certain types of NFRs is low.

2.1 Analysis

Some opportunities motivated us to conduct our research:

- Most of the related works have used the PROMISE dataset, which has at least two issues: (i) it is imbalanced, and (ii) the number of NFRs is not numerous (370 NFRs). For our investigation, we combined two datasets (PROMISE and PURE (with 160 NFRs)) totalling more than 500 NFRs. In addition, we separately apply (or preliminarily validate) our techniques on an industry dataset (of approx. 260 NFRs).
- We cover a number of techniques in combinations: three feature extraction techniques (BOW, TF-IDF, and POS-tagging) and 10 supervised machine learning techniques (SVM, SGD SVM, DT, Extra Tree, Bagging Tree, LR, RF, GNB, BNB, and MNB).

3 Research Investigation

As mentioned in Sect. 1, our investigation focuses on four NFR types: usability, performance, security, and operational NFRs. The key research question we ask is:

Which combination of feature extraction techniques (BOW, TF-IDF, and POS-tagging) *and SML algorithms* (SVM, SGD SVM, DT, Extra Tree, Bagging Tree, LR, RF, GNB, BNB, and MNB) *yields the highest precision and recall values?*

The higher the precision value, more the proportion of the identified NFRs that are relevant to the inquiring stakeholder. The higher the recall value, less the relevant NFRs that are missed (or not identified) from the dataset.

In a practical setting, the best combination of feature extraction technique and the SML algorithm could be a basis for a real-time interactive tool that serves the needs of various kinds of stakeholders: analysts, architects, programmers, testers, product managers, domain experts and more. They all need to know the details of the NFRs relevant to their concern, in a given project, from time to time. For example: What are the reliability needs of the system and have I addressed them all in the design of the system's architecture? Is my choice of algorithms to code appropriate for the performance requirements of this device? Does the envisaged core quality of the system give us a competitive advantage?

3.1 Datasets

We use two datasets: NFR PROMISE dataset and the PURE dataset[1]. The PROMISE dataset consists of 625 requirements (255 functional, 370 NFRs). There are 11 groups of NFRs but because the number of instances of a few types is low, we chose the top four types for classification: usability (62), performance (48), security (58), and operational (61). We also used the PURE dataset (which focuses on public requirements documents). It consists of 296 requirements (136 functional and 160 NFR). There are 16 groups of NFRs in this dataset and the top three types are usability (54), performance (18), and security (17).

---

[1] https://zenodo.org/record/1414117#.X7wfOWhKiUk.

We combine these two datasets in a sequence to form the training set. The effect on precision and recall of mixing the two datasets differently (e.g., interleaved NFRs) is not empirically tested though it seems that the SML algorithms are agnostic about the ordering of the NFRs in the dataset.

Given a dataset of all requirements, we first identify functional requirements and, separately, NFRs. For this, we use the distinguishing criteria from the established literature [5]. We then further classify the NFRs into the four (for PROMISE) and three (for PURE) predominant types. By stripping off extraneous metadata, we then convert the datasets into standard CSV files. They include the columns: "Requirement Description", "F/NF" for functional or NFR, and "Subtype" of NFRs. Table 1 depicts example NFRs from the PROMISE and PURE datasets.

Table 1. Example NFRs from the PROMISE and PURE datasets

Label	Requirement text	Dataset
Availability	"Aside from server failure the software product shall achieve 99.9% up time"	PROMISE
Safety	"The system will do periodic backups through a live internet connection"	PURE

3.2 Research Methodology

We want to determine the best combination pair of the (i) feature extraction and (ii) supervised machine learning, techniques. We overview this process in this Preview paper; for further details see (the first author's thesis): https://ir.lib.uwo.ca.

Step 1: Data Preprocessing
In this step, we first parse the CSV file into a DataFrame for the convenience of processing with Python, with each row in the DataFrame representing a single requirement sample. This involves: (i) parsing the given csv file, (ii) using NLTK library to remove case distinctions, (iii) tokenization and punctuation removal, (iv) stop word removal (v) part of speech tagging, and (vi) stemming and lemmatization.

Step 2: Feature Extraction
We now want to extract features as input to our classification algorithm:

- Requirements are converted into numeric vectors using the BoW in [6].
- TF-IDF scores associated with each term present in a given requirement is used in this classification framework.
- We adopted a feature list proposed by Hussain et al. [7] with which they attained 95% accuracy in the binary classification of FRs and NFRs. (However, note that they did not classify NFRs into sub-types) Thus, adopting the NFR-characteristics from [5], we added a number of syntactic features (e.g., # Adjectives, #Adverbs, and #Nouns) and coded keyword features as part of speech groups.

At this point, three feature sets are prepared. For each of them we have xtrain (data frame with requirement features) and ytrain (data frame that includes target values): (i) For the classification of functional/NFR, ytrain contains 1 (for functional) or 0 (for NFR); and (ii) For multiclass classification of NFRs, ytrain contains 1 for usability, 2 for security, 3 for performance, and 4 for operational types. The output of this step is the three extracted feature sets that are input to machine learning algorithms of training classifiers.

Step 3: Training Classifiers

Here, we investigate the performance of the 10 mentioned classifiers using stratified 10-fold cross-validation technique [2]. Each time the dataset is divided into 10 subsets, nine are used for training and the remaining one is used for testing. We repeat the process 10 times and the performance of the classification is measured as average precision and recall of the 10 repetitions complemented by their execution time.

Step 4: Classifying Requirements

Each of the classifiers trained in Step 3 is used to predict for each requirement whether it belongs to usability, performance, security, or operational.

4 Preliminary Evaluation

This section describes the results of the described investigations. In Sect. 4.1, we give precision and recall to evaluate how well the model learnt to classify non-functional requirements. Comparison with related work is also described. In Sect. 4.2, we show the performance of the model using an industry dataset. Results of the experiments with this dataset are shown as well.

4.1 Preliminary Analysis

Table 2 gives an overview of the results of multi-class classification of NFRs. The average of POS, BoW, and TF-IDF shows that SVM achieved the best results with recall of 0.88 and precision of 0.89. LR, Extra Tree, MNB and SGD- SVM with recall values of 0.86, 0.85, 0.85, and 0.84 (resp.) performed well too. All classifiers achieved recall values above 0.8 except DT, Bagging Tree, and BNB.

Among all the combinations of feature extraction techniques and machine learning algorithms, SVM with TF-IDF (recall: 0.9, precision: 0.92), LR with POS and BOW (recall:0.90, precision: 0.87), and MNB with BOW (recall: 0.90, precision: 0.88) achieved the best results. However, further empirical work is needed to assess the root-causes of the 70-odd misclassified NFRs and how to improve the performance.

Comparison with related work, e.g. [2, 4], is generally favourable (except DT (0.74) and Bagging Tree (0.75) vs. [2]: 0.76). Our best case SVM/TF-IDF vs. Related work are depicted in Table 3.

Table 2. Results of NFR classification

Algorithm	POS			BOW			TF-IDF			Averege		
	Precision	Recall	Time	Precision	Recall	Time	Precision	Recall	Time	Precision	Recall	Time
SVM	0.89	0.87	0.13	0.88	0.87	0.1	0.92	0.9	0.08	0.89	0.88	0.1
SGD SVM	0.85	0.83	0.25	0.88	0.86	0.27	0.84	0.82	0.3	0.86	0.84	0.27
LR	0.9	0.87	0.09	0.90	0.87	0.9	0.9	0.85	0.09	0.9	0.86	0.09
DT	0.78	0.76	0.14	0.77	0.74	0.16	0.76	0.72	0.17	0.77	0.74	0.16
Extra tree	0.88	0.85	1.6	0.88	0.84	1.72	0.9	0.86	1.8	0.89	0.85	1.7
Bagging tree	0.81	0.76	0.87	0.8	0.76	0.94	0.81	0.74	1.03	0.81	0.75	0.94
RF	0.82	0.78	0.2	0.85	0.81	0.22	0.84	0.8	0.21	0.84	0.8	0.21
GNB	0.81	0.81	0.08	0.82	0.82	0.1	0.8	0.79	0.1	0.81	0.81	0.09
MNB	0.88	0.86	0.04	0.9	0.88	0.05	0.89	0.83	0.05	0.89	0.85	0.04
BNB	0.85	0.77	0.08	0.87	0.75	0.1	0.88	0.78	0.11	0.87	0.77	0.09

Table 3. Best case comparison with related work

	Precision	Recall	Execution time
Our	0.92	0.90	0.08s
Reference [2]	0.24	0.76	Not specified
Reference [4]	0.71	0.72	Not specified

4.2 Preliminary Validation

Table 4 shows recall values ≥0.89 and precision values ≥0.92 of the three classifiers using an industry dataset (262 NFRs): SVM, LR, and MNB, for multi-class classification of the four mentioned NFR types. An example NFR from this dataset is:

"The xxx shall perform a failover between an active xxx node and its redundant xxx node partner in no more than 5 seconds."

Table 4. Multi-class NFR classification of the industry dataset

Algorithm	POS			BOW			TF-IDF		
	Precision	Recall	Time	Precision	Recall	Time	Precision	Recall	Time
SVM	0.93	0.91	0.09	0.96	0.95	0.06	0.96	0.96	0.09
LR	0.93	0.92	0.03	0.96	0.94	0.04	0.93	0.90	0.04
MNB	0.92	0.90	0.02	0.94	0.9	0.03	0.92	0.89	0.02

5   Conclusion and Future Work

We used three feature extraction techniques: BoW, TF-IDF, and POS-tagging, combined with 10 supervised machine learning algorithms SVM, SGD SVM, LR, DT, Extra Tree, Bagging Tree, RF, GNB, MNB, and BNB for classifying four NFR types (usability, performance, security, and operational NFRs). Also, we used a combination of PROMISE and PURE datasets (see Sect. 3.1) as our training set. SVM with TF-IDF, LR with POS and BOW, and MNB with BOW achieved the best results (see Table 2) with recall value over 0.90 and precision value over 0.87.

Comparison with related work is favourable (see Sect. 4.1). We also preliminarily validated our results, using an industry dataset (see Table 4), showing recall values ≥ 0.89 and precision values ≥ 0.92. These values suggest that the right NFRs would be rendered, and quickly, to the stakeholders concerned.

In this research, we used a multi-class classifier for the specified NFR types. For future work, we intend to investigate whether developing binary classifiers for each of these NFR types could possibly improve the performance of the classification task at hand.

Also, in our research thus far, we have considered product quality attributes such as usability, performance, security, etc. Apart from extending the coverage of the product quality attributes (e.g., reliability, privacy, and security), a further step in providing a concern-based NFR classification tool for stakeholders is to consider *process* quality aspects such as implementation risk, effort, and cost, which are the bedrock of software projects in industry. For example, recognising risky requirements can help in prioritising these requirements over less risky ones in early decision-making in the development process using the Spiral model of development [8].

Acknowledgments. This work is supported, in part, by Natural Science and Engineering Research Council (NSERC) of Canada.

References

1. Nekvi, I., Madhavji, N.H.: Impediments to requirements-compliance in contractual systems engineering projects: a case study. ACM Trans. Manage. Inf. Syst. **5**(3), 15, 1–35 (2014)
2. Cleland-Huang, J., Settimi, R., Zou, X., Solc, P.: Automated classification of non-functional requirements. Requirements Eng. **12**(2), 103–120 (2007). https://doi.org/10.1007/s00766-007-0045-1
3. Zhang, W., Yang, Y., Wang, Q., Shu, F.: An empirical study on classification of non-functional requirements. In: Proceedings of the 23rd International Conference on Software Engineering and Knowledge Engineering (SEKE), pp. 190–195 (2011)
4. Lu, M., Liang, P.: Automatic classification of non-functional requirements from augmented app user reviews. In: Proceedings of the 21st Int. Conference on Evaluation and Assessment in Software Engineering, June 2017, pp. 344–353 (2017)
5. Sommerville, I., Sawyer, P.: Requirements Engineering: A Good Practice Guide. Wiley, New York (1997)
6. Slankas, J., Williams, L.: Automated extraction of non-functional requirements in avail able documentation. In: Proceedings IEEE 1st Workshop NaturaLiSE, pp. 9–16 (2013)

7. Hussain, I., Kosseim, L., Ormandjieva, O.: Using linguistic knowledge to classify non-functional requirements in SRS documents. In: Kapetanios, E., Sugumaran, V., Spiliopoulou, M. (eds.) NLDB 2008. LNCS, vol. 5039, pp. 287–298. Springer, Heidelberg (2008). https://doi.org/10.1007/978-3-540-69858-6_28
8. Boehm, B.W.: A spiral model of software development and enhancement. IEEE Comput. **21**(5), 61–72 (1988)

RE for AI-Enabled Systems

AdaptationExplore – A Process for Elicitation, Negotiation, and Documentation of Adaptive Requirements

Fabian Kneer[1]([⊠]), Erik Kamsties[1], and Klaus Schmid[2]

[1] Dortmund University of Applied Sciences and Arts,
Emil-Figge-Str. 42, 44227 Dortmund, Germany
{fabian.kneer,erik.kamsties}@fh-dortmund.de
[2] University of Hildesheim, Universitätsplatz 1, 31141 Hildesheim, Germany
schmid@sse.uni-hildesheim.de

Abstract. [**Context and motivation**] Current and future systems have to operate in complex and dynamic environments. An adaptive system addresses these challenges as it monitors its environment and reacts by changing its behavior. [**Question/Problem**] Representations of adaptive requirements (e.g., at runtime) and strategies for decision-making have gained a lot of interest in past and current research. Yet, there is a lack of support for *elicitation* of requirements and environmental information for adaptive systems.

[**Principal ideas/results**] We suggest to apply creativity techniques to elicit *adaptation* requirements and make use of *situations* to negotiate them (a situation represents the state of the system and its environment at a particular instance of time). [**Contributions**] In this paper, we introduce *AdaptationExplore*, a process for the development of adaptive systems, which supports engineers in particular during the early phases. The results of a pilot study are reported. 37 Master students applied the process on different cases. The study provides first positive experiences on the effectiveness and applicability of the process.

1 Introduction

The notion of adaptive systems, i.e., systems that react to observations and adapt their behavior accordingly has gained significant interest in the research community, due to its wide applicability [1]. The part of the real world that is relevant to understand a system is usually called its *environment* (while the term *context* is used in the domain of adaptive systems as well, we stick to *environment* in this paper). A proper understanding of the environment is a prerequisite to write requirements for a system. A requirements engineer needs information and knowledge about the environment to identify and develop possible adaptations of a system.

An adaptation undertaken by an adaptive system is usually due to changes in the environment. While the environment of a system always needs to be taken

© Springer Nature Switzerland AG 2021
F. Dalpiaz and P. Spoletini (Eds.): REFSQ 2021, LNCS 12685, pp. 81–98, 2021.
https://doi.org/10.1007/978-3-030-73128-1_6

into account in engineering, in developing adaptive systems, there is a need to provide the system itself with a formalized notion of an environment model.

In this paper we introduce *AdaptationExplore* – a RE process for elicitation, negotiation, and documentation of requirements for adaptive systems. The goal of our work is to show that creativity techniques can be beneficially used to identify adaptation requirements. In order to do so, we introduce the notion of situation and create a specific process to demonstrate the feasibility of this goal. The contribution lies in the coverage of the early phases of the development of an adaptive system. The process embodies several novel ideas:

- *Creativity techniques* are helpful in generating questions [2], we employ trigger questions to explore the environment to **elicit** requirements. Besides the work by Dey and Lee [3], the application of creativity techniques in the development of adaptive systems was not described in the literature so far, to the best of our knowledge.
- We introduce the concept of a *situation*, commonly understood as something like "the set of conditions that exist at a particular time in a particular place", to the development of adaptive systems. Our idea is that looking at situations *slices* problem and solution domain in a way that is beneficial to the development of adaptive systems as it helps to deal with the complexity that arises from the dynamics of the real world. In particular, situations make it easier to uncover especially those requirements, which are adaptation-relevant, when situations are integrated and conflicting requirements are **negotiated**.
- A lightweight *tabular goal model* is used for **documenting** requirements. A goal model has its strengths in modeling hierarchies, dependencies, and different goal realizations. We suggest a simplified goal model, which consists only of goals, tasks, resources, and dependencies. We omit a means to model *different* goal realizations (as the "mean-ends" link in i*), as a situation in our view has a goal but only *one* realization. The simplifications allow us to use a *tabular* representation, which makes a goal model more concise.

The remainder of the paper is structured as follows. The related work is discussed in the following section. In Sect. 3, an overview of the process is provided. Section 4 introduces a running example that is used to describe the process. The three phases of the process are explained in detail in Sect. 5, 6, and 7. A pilot study on *AdaptationExplore* with Master students is presented in Sect. 9. Finally, we conclude the paper and discuss the future work in Sect. 10.

2   Related Work

Adaptive systems have gained a lot of interest in RE research. The challenges of requirements and decision making for adaptive systems have been noted frequently. Different representations are proposed like goal models (KAOS, i*, etc.) [4–9]. Decision processes based on various formalisms were suggested, for example, rule based systems [10] and dynamic decision networks [11].

Modeling and understanding the context of a system is an important part of developing an adaptive system. The need for enhanced requirements elicitation and a methodology that supports all stages of the development of context-aware systems was identified by Alegre et al. [12].

Kolos-Mazuryk et al. [13] compared three goal modeling methods, i*, KAOS, and GBRAM. The authors concluded that all three methods lack capabilities to explicitly model the system environment. Paja et al. [14] have shown that goal modeling techniques are a good starting point for strategic decision making due to the global view on the problem. In their experiment, Business Intelligence Model (BIM) performed better than i*, because of the integration of context information, which enhance the decision making process.

Soares et al. [15] compared three goal modeling techniques - Tropos4AS, AdaptiveRML, and Design Goal Models. They tested if the approaches are suitable to cover a set of 20 modeling dimensions of adaptive systems. All three techniques performed well in modeling goals, changes, and effects of an adaptation. A lack was identified in regard to mechanisms, that is the system's reactions towards changes in terms of e.g., scope, duration, timeliness, and triggering.

We identified three approaches in literature that propose a *process* for developing adaptive systems. Cheng et al. [16] use goal modeling (KAOS) and a variation of threat modeling to identify requirements. Cheng uses a goal tree and a conceptual domain model as a structure to follow in order to identify uncertainties. The idea is a top-down and bottom-up analysis of the goal tree. Top-down a goal hierarchy with sub-goals is built. Uncertainties or environment information, which have effects on the satisfaction of the goals, are added at the bottom of the tree. The mitigation of uncertainties take place in the bottom up analysis. Here, possible problems will be solved by strategies like adding high-level goals, sub-goals, or relaxing a requirement.

Another goal-based process was introduced by Morandini et al. [17]. They developed Tropos4AS, an extension to the Tropos goal modeling framework to model context, conditions, and errors. The idea is to develop a normal Tropos goal model and then to extend it by environment information and conditions that describe influences on the goal satisfaction. The next step is the identification of failures. Here, possible failures and their reasons are identified. A failure model is added, which describes how to solve the error by adapting the system.

These processes describe how to develop adaptive requirements, but it was observed, that the *elicitation* phase, which is part of every traditional RE effort, is not explicitly covered, for example by [3,18]. Instead the processes offer some kind of *systematic exploration* of the environment in absence of an elicitation phase.

Dey and Lee [3] use a cognitive technique called *Repertory Grid* in combination with problem and design space exploration. The idea is to identify conflicts between features in the different contexts in the problem space and resolve them. After a set of problem spaces was identified the design space is explored by adding design decisions and identify aspects that have effects on the design space. Finally, the design space is filtered by user preferences.

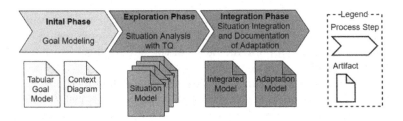

Fig. 1. Overview of the *AdaptationExplore* process

Quite a few elicitation techniques are known in RE, but most of them are of little help as adaptivity is related to uncertainties ("known unknowns" and "unknown unknowns"). Elicitation techniques primarily help in *gathering information* to answer questions rather than help to raise the *right questions*.

This is the point where *AdaptationExplore* comes into play. One contribution of the process is to employ creativity techniques to empower engineers to become more innovative in the elicitation phase and discover potentially relevant aspects.

Creativity technique is a rather broad term, encompassing simple approaches such as checklists (e.g., Osborn's checklist) as well as complex methods like TRIZ.[1] To foster creativity, we use so called *trigger questions*. A simple form of which is W5H1 (Who, Why, What, Where, When, and How). More advanced triggers are SCAMMPERR [19] or the trigger list by Robertson and Robertson [20]. They focus on triggering creativity via specific customer aspects or qualities, for example "speed - what can you make faster?", "connectivity - what new connections can you make?". New triggers were added for example "entertaining - add a feature that make it fun to use the system" by Burnay et al. [21]. Our work builds previous approaches and can be viewed as a first attempt to design creativity triggers for the development of *adaptive* systems.

3   Process Overview

The *AdaptationExplore* process was developed following a design science approach [22]. The goal is to build a continuous requirements engineering process for the development of adaptive systems. In a first iteration, we explored the notion of the environment. We identified different modeling approaches for the environment, but also identified a lack of elicitation approaches for environment information. This leads to a second iteration of *AdaptationExplore*, which targets at a better understanding of the environment with the help of situations and creativity triggers to enhance the elicitation phase, in particular the identification of adaptation scenarios.

AdaptationExplore in its current shape consists of three phases named *Initial Phase*, *Exploration Phase*, and *Integration Phase*. Figure 1 shows the phases with the artifacts they produce.

---

[1] https://www.mycoted.com/Category:Creativity_Techniques.

In the **Initial Phase**, first information about the system is gathered with the help of stakeholders and domain experts. The objective is to define goals, features, and resources for the adaptive system. The results are a tabular goal model and a context diagram.

The **Exploration Phase** uses the tabular goal model to identify possible situations of the system and the environment. For every situation, an information and a behavior model is developed. During exploration, the different situations are analyzed with the help of a questionnaire which contains 15 trigger questions (TQ). The questions guide the developer to spot missing environmental information, to identify new situations, and to reduce the uncertainties in a situation. The exploration phase leads to new insights for the initial phase and to a refinement of situations. The result of the exploration phase is a set of situations (with accompanying information and behavior models).

In the **Integration Phase**, the different situation models are combined into a joint information and behavioral model. The integration of models typically leads to conflicts. A conflict is an indicator for a required adaptation between one or more situations. The objective of the third phase is to solve all conflicts and to specify adaptation points. The following sections present the three phases in detail using a running example.

4 Running Example

We use an alarm clock as our running example to illustrate the details of *AdaptationExplore*. Our alarm clock is assumed to be an app running on a smart phone. On the main screen of the app the clock displays the current time and the alarm time. A configuration screen is used to manage the alarm time and to turn the alarm clock on and off. The main task is to wake up the user at a pre-defined time (alarm time). Besides the time, the user can also specify the days of a week when the alarm clock should wake her up, e.g. only Monday to Friday. The alarm sound can be changed by the user, she can choose between a set of predefined sounds from the smart phone. When an alarm starts, the sound is played, the smart phone starts to vibrate, and a stop button is shown on the display.

A snooze function is used to stop the current alarm and shift it by a time given by the user, e.g. 5 min. If the snooze function is activated in the configuration view, a snooze button will be shown next to the stop button if an alarm starts.

5 Initial Phase

Figure 2 depicts the initial phase of *AdaptationExplore*. The goal of the initial phase is to gather a first overview of the system and its environment by identifying goals, tasks, resources, and the relations between them. The goal modeling activity results in two models: (a) a goal model that captures and refines the goals, defines tasks, provides alternative adaptations to satisfy a goal and resources that are needed; (b) a context diagram, which captures information

Fig. 2. Initial phase of the *AdaptationExplore* process

(R4) Traffic Information Provider	(R3) Calendar	(R2) Smartphone	(R1) User	(G1) Maximize Sleep Time		+	---
				(G2) Easy to Use	--	+	+	+++
				(G3) Wake Up on Time	+		++	---
				Task/ Functionality	(T1) MI	(T2) CI	(T3) TII	(T4) Base
				Data In/out				
			1	Alarm time				x
			1	Snooze time	x	x	x	
			1	Setup time	x	x	x	
2			1	Journey duration	x	x		
			1	Arrival setup time	x	x	x	
			1	Buffer	x	x	x	
	2		1	Arrival time	x		x	
	2		1	Destination	x		x	
		2	1	Current position	x		x	
		1		Internet connectivity			x	

Fig. 3. Tabular goal model illustrating goals, tasks/functionalities, and resources - MI Manual Import, CI - Calendar Import, TII - Traffic Information Import

about the environment. The benefits of goal models and context diagrams in the early stages of development was already shown by Cheng et al. [16] and [17].

We propose a *tabular goal model* to establish a pragmatic starting point and to assist later in the integration of new elements during the exploration phase. The tabular goal model is a simplification of i^* for the purposes of our process, it embodies a subset of concepts and brings them into a tabular representation. The main differences are: we do not separate between hard and soft goals, and we left out the concept of actors and their boundaries. For the later stages of our process, just goals, tasks (*functionalities* from a runtime perspective), resources, and dependencies are relevant. Our process is not tied to tabular goal models, any goal modeling approach would work with our process, for example KAOS or Tropos [16,17].

An example of a tabular goal model is provided in Fig. 3. The example contains three goals (G1–G3): the user wants to sleep as long as possible, i.e., maximize sleep (G1), but also wants to wake up on time (G3). Further, the alarm clock should be easy to use (G2).

Four tasks (T1–T4) were identified. Besides the base task (T4), which was already described in Sect. 4, these are:

(T1) Manual Input: Optionally, the user manually enters the information needed to calculate the optimal alarm time.

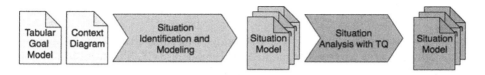

Fig. 4. Exploration phase of the *AdaptationExplore* process

(T2) Calendar Import: The user can choose a dynamic alarm clock to import information from the calendar app to set the alarm time. The start time of the first appointment in the morning and travel time to the its location is used to calculate the alarm time.

(T3) Travel Information import: This task assumes a connection to a navigation system to calculate the time needed to travel to the destination and adjust the alarm time accordingly. This means, if there is a traffic jam on the way to the office, the alarm time is shifted based on the needed time to drive to the office. The according functionality at runtime is only available if the GPS is active on the smart phone, the location of the first appointment is known, and a connection to the navigation system is present.

The table shows in the upper-right matrix, formed by tasks and goals, the positive or negative impact (--- to +++) that a particular task has on a given goal. This resembles contribution links in the i* notation.

The Data in/out part of the table lists the relevant data items for the system and the lower-right matrix relates those data items to the tasks. A cross in a particular cell denotes that the task uses this data item as an input or output.

The entries R1–R4 denote the different resources that can be used by the system. As we do not model the system boundaries or connected actors, a resource can also represent an actor (e.g., the user), who provides or needs data. (R1) The user is the first resource for information. As the application runs on a (R2) smartphone, smartphone applications like (R3) a calendar or a traffic information provider (R4) can also be used.

In this example the user is the main resource for information. Only if he is not available the second source should be used. This is shown in the priority matrix in the lower left part. It relates data items to resources and shows the ranked priority between a data item and from which resource it should be taken.

6   Exploration Phase

The exploration phase aims at unfolding previously unknown features of the environment. For this purpose specific situations of the system and the environment are modeled and analyzed to understand the behavior of the system and the involved environmental entities. Figure 4 shows the steps and results of the exploration phase. Core concept of this phase is a so-called *situation*.

6.1   Situation

The term *situation* is commonly understood as something like "the set of conditions that exist at a particular time in a particular place"[2]. The term has also a distinct meaning in philosophy and psychology. In short, it has a *perspective*, it *frames the real world*, and it has an often implicit *subject*. We believe the conception behind the term is useful in developing adaptive systems as it helps to deal with the complexity that arises from the real world. Looking at situations *slices* problem and solution domain in a way that is beneficial to the development of adaptive systems. In the remainder of this subsection, we define the term in the context of adaptive systems and compare it to related terms such as *scenario*.

We define a *situation* as an abstraction that embodies the structure, function, and behavior of a system and the relevant part of the environment at a particular location over a time period of particular interest.

The concept of *views* and *viewpoints* has gained a lot of interest in the past, inside RE e.g., [23] and in SE in general. A characterizing element of a view is a viewpoint, for instance taken by an actor or stakeholder. Similar to a view, a situation can be incomplete. In contrast, in a situation, the notions of environment, location, and period of time are more prevalent.

The term *scenario* stands for a broad range of related concepts. For example the NASA [24] use *operational scenarios* to develop dynamic views of the systems operations. The idea is to analyze a function in various modes and mode transitions. Also included in the analysis is the interaction of the system with external interfaces and the reaction on faults and errors. Sutcliffe [25] has summarized advantages and disadvantages of scenarios. A major advantage is that scenarios capture patterns of the real world. Another advantage is the level of detail and because of this, they are easy to understand also by non-technicians. The disadvantages are related to the lack abstraction and their quantity. A situation is the basis for a creative process to identify potential scenarios. In other words, a situation is a crystallization point for unfolding adaptation scenarios. From a scenario perspective, a situation is a starting point for an adaptation scenario sequence.

In conclusion, a situation is most closely related to context among the discussed terms and is an attempt to make it more precise in a way that furthers the creative process. The strength of situations are their ability to slice a description in a way that fits to adaptive systems, the quantity of possible situations is a clear drawback.

6.2   Identification of Situations

The first step is to identify situations, the goal is to identify different realizations of a task in a given situation. By focusing on a specific situation, missing alternatives or required resources can be identified. The first situations are identified using the goal and context model of the initial phase. The result is a set

---

[2] https://www.macmillandictionary.com.

of situations. Each situation is defined by a description, an information model, and a behavior model.

Textual Description. We suggest a textual description of a situation. The following form contains an example of a situation describing the *base* task. The form helps to frame a situation that should be analyzed in more depth.

---

1) Situation Name: "Wake person in bedroom (Base)"
2) Information Source: "Martin Spencer"
3) Description: "The user is in his bedroom using the alarm clock to wake up on time. The
↪ alarm clock is set up to wake the user at 8 o'clock with a possible delay by the snooze
↪ function."
4) Task: "Base"
5) Environment: "Smart phone, user"
6) System: "checkAlarm(), snooze(), alarm()"
7) Constraints:
   Environmental: "-"   Non-functional: "-"

The form gathers general information of a situation such as its name (1), source (2), for example a stakeholder or document, and a short description (3). The task slot (4) refers to the name of the task that is analyzed and described in the situation (with a focus on a single realization, no adaptation), the base task of the alarm clock in the running example. The environment slot (5) describes when/where/what environment elements (connected devices, user, etc.) are available. The system slot (6) contains information about the active elements of the situation model (e.g., available methods, internal components) used in the situation. Finally, the form captures constraints (7) in the environment or non-functional constraints that we can not influence or change. In the example situation, we have not much information about the environment and we need to investigate by modeling the situation and analyze it.

Situation Modeling. For the supplementary modeling of a situation we suggest two models, an *information model* (e.g., UML class diagrams) and a *behavior model* (e.g., UML activity, sequence, or statemachine diagram). Each aspect that is relevant for the task/functionality and all environmental elements that are available in a situation are to be covered by these models.

We start with the situation "Wake person in bedroom (Base)" that is used to analyze the base task of the alarm clock. Figure 5 shows the information model for the situation. All identified entities are modeled in the active context, which means we need to analyze this element in the next step. The model contains the alarm clock with the attributes and methods that are needed for the base task. It also contains first attributes gathered for the smart phone and a class for the user. The situation models are used next to identify tasks that are not sufficiently understood.

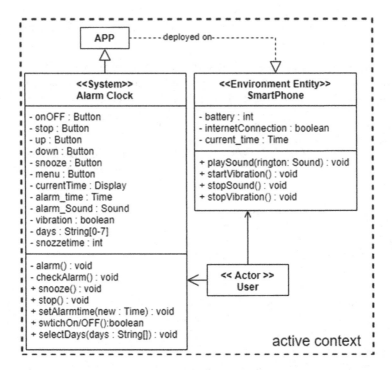

Fig. 5. Information model for base task

6.3   Situation Analysis with Trigger Questions

The analysis is guided by trigger questions (TQ). We selected trigger questions as a lightweight creativity technique. The questions were developed by using the 5W1H (Kipling) technique as a basis. The idea was to identify questions that could trigger new ideas for missing environment entities, tasks, alternative realizations, situations, or data items. We started with the general question of 5WH1 and derived from them 15 trigger questions, see Table 1.

The first four questions aim at the highest abstraction level and should be used on classes, entities, and similar high-level elements of the situation model. The next eight questions are related to methods and the effects on the system and its behavior, needed resources, alternative realizations. The final three questions support the analysis of variables of a situation. They try to identify influences, resources, or calculations for a variable. The questions should be answered for each element in a model to identify new elements and to systematically analyze the situation. New ideas are triggered with a question, but additional considerations are needed. For example, the question "Could the method fail?" can be answered with yes or no, but it directs the attention to an important topic, which could lead to new ideas for adaptations or triggers for an adaptation.

When we start with the user and ask the question "What should the model element be related to?", we could add the structure of the environment by adding

Table 1. *AdaptationExplore* trigger questions

(1) What should the model element be related to?	(9) What information should the method provide and for what could the information be used?
(2) Does the model element have missing restrictions?	(10) How could the method be accessed?
(3) What information could the model element provide for a feature and how could the model element be used by a feature?	(11) Could the method fail?
(4) How should the model element be accessed?	(12) Could there be any reasons not to perform the method?
(5) On what should the success of the method depend on?	(13) What influences the variable?
(6) What should the method be related to?	(14) Where could the value(s) of the variable come from?
(7) Does the method have missing restrictions?	(15) What is the variable used for?
(8) Could there be any obstructions or conflicts due to other methods?	

a building, which consists of rooms. In a room we can have furniture and smart devices. These elements are high level abstractions that represent the environment and will be added to the *passive context* that contains all elements, which do not need further analysis or which are already analyzed.

When the process goes on and we analyze the smart phone and look at the question "What should the model element be related to?", we identify a smart device that is connected to the smart phone that is only available in some rooms. The element is a *smart light* that can be controlled by switching the intensity of the light (see green entity in Fig. 6). The light is not relevant for the current situation so it is modeled in the passive context. For the light we identify a new task *natural sunrise wakeup*. This task is added to the goal model as well as the resource *smart light*. Furthermore, a new form is added to describe the situation *"wakeup by natural sunrise in bedroom at home (light)"*.

(T5) Natural sunrise: A situation was identified in which a user owns a smart lamp that can be controlled and dimmed via the smart phone. Waking up the user is made more pleasant by slowly dimming the lamp before the alarm starts.

The result of the situation analysis is a set of analyzed situations that represent the tasks in different situations and show the alternative realizations based on the given environment in a situation.

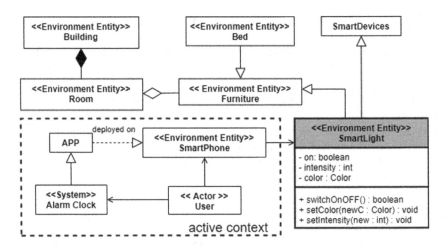

Fig. 6. New environment entities for the base task

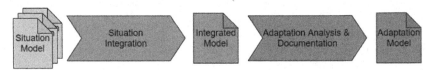

Fig. 7. Integration phase of the *AdaptationExplore* process

7   Integration Phase

The integration phase is shown in Fig. 7. In the integration phase the developed situations are merged to identify differences, like alternative realizations of tasks, different resources, attributes, or attribute values. Differences are analyzed to identify whether adaptations are needed to solve conflicts between situations. If a need for an adaptation is identified, it is described by an adaptation point. An adaptation point is a location in the system where we need to solve a conflict between different situations and the related realization of a task. A change could be to switch between different realizations of a task, to enable/disable functionalities, to change a resource or an attribute. The result of the phase is an adaptation model.

7.1   Situation Integration

First, the information models are integrated into a joint model. During integration, conflicts between models may become apparent. A conflict could be a different input source, alternative realizations of a task, or other ways to provide the results of a task. Furthermore, if an external element was identified in a situation that is only available in this specific situation it also indicates a need of an adaptation and must be investigated, for example the availability of the smart light for the task natural sunrise.

NR	Start		Changes in the System			Limitations		Reachable Situations
	Situations	Changes to Situation	Status Changes of Functionality	Functionality Changes	Variable Changes	Time	Location	
1	Base, Light	Connection to Traffic Provider and GPS	enable TII()	checkAlarmTime() using TII()	-	-	-	Travel
2	Base, Travel	connected to Smart Light	enable naturalSunrise()	checkAlarmTime() using ns()	Alarm -30 min	Alarm -31 min	-	Light
3	Light	connection lost to Smart Light	disable ns()	checkAlarmTime() not using ns()	Alarm +30 min	Alarm -31 min	-	Base, Travel
4	Travel	No internet connection	disable TII()	checkAlarmTime() using default value	-	-	-	Base, Light

Fig. 8. Example of an adaptation model for the alarm clock

A conflict can be an indicator that an adaptation is needed, but it can also result from an ordinary inconsistency. Also, adaptations add complexity to a system and therefore introduce additional costs. That is, a decision for one alternative at development time might be the preferable option. A human judgement is required in any case.

Next, the behavior models are integrated. In particular the integration of behavior models may result in an oversized joint model. In this case, a solution is to integrate behavior models in a task-wise fashion, so that one joint model per task results. Conflicts that can not be solved during the integration of for example sequence diagrams could be marked by adding an additional element, named adaptation framework, which is used to decide which behavior should be used.

7.2 Adaptation Analysis and Documentation

First, conflicts and inconsistencies remaining from the previous step are resolved by introducing adaptation points. The concept of adaptation points is based on the notion of variation points of software product lines (SPL), in particular of dynamic software product lines (DSPL) [26,27]. In a DSPL the adaptations are described by variations points, which are organized in a variability model.

Identification of Adaptation Points. Conflicts between task realizations could indicate an adaptation point, e.g.:

- different sampling rates of sensors,
- errors in hardware or software components that are required for performing a functionality,
- availability of elements in the environment, e.g. connected systems or actors, which are only available at a specific time or place,
- conflicts between goals due to different realizations of tasks.

Next, the identified adaptation points are investigated by looking at differences of the involved situation models (of two or more situations). A so-called *adaptation model* (see Fig. 8) describes the necessary changes to address these differences. It defines the model changes in the transitions among the three situations *Base*, *Light*, and *Travel*. The Base situation describes the base task of the alarm clock. In the Travel situation the user has an Internet connection and calculates the journey duration via Internet to compensate, for example, a traffic jam. The main task is *TII - travel information import*. The Light situation requires a smart light in the environment to enable the task *ns - natural sunrise*.

The *Start* (second column) of an adaptation is a set of situations, for example in the first row the situations Base and Light. Start also includes information about a change in the situation that leads to an adaptation. In this case the connection to the *traffic information provider* is established.

The third column describes what changes are made in the system, this includes activating or deactivating functionalities, internal changes to functionalities, or changes to variables. In the first row the change is enabling the *TII()* functionality, which is used in the *checkAlarmTime* functionality to calculate the journey duration. In the second row, which describes the adaptation to the situation Light, the variable *alarm* needs to be changed as dimming of the smart light should start 30 min before the alarm starts.

The limitations (fourth column) help to minimize errors produced by adaptation, for example to wait for a running process, or to finish before the adaptation can take place. For the adaptation in the second row we have defined a limitation, it can only happen 31 min before the alarm starts to ensure the user wakes up on time.

Finally, an adaptation point is marked in the information and behavior model to show the location where a specific adaptation takes place or in the form of a list with a textual description. At these points a decision is needed and the system may switch to another behavior. The identified criteria for the decision process are collected in the list and linked to adaptation points.

8 Discussion

Similar to other goal-oriented approaches for adaptive systems like Cheng et al. [16] and Morandini et al. [17], *AdaptationExplore* also starts with a goal model. However, *AdaptationExplore* tries to cope with the complexity of identifying adaptive requirements by analyzing one situation at a time. We assume that this will simplify the creative task. The idea is to understand a situation and to find the best realization in this specific situation without thinking about other situations and the adaptations relevant to them. The environment is in the focus of analysis and investigating a single situation often leads to the identification of new aspects of the environment and also new realizations to cope with an environmental aspect. Goal-oriented processes on the other hand deal with the whole domain model or the goal hierarchy.

Similar to problem space exploration as suggested by Dey and Lee [3], we want to identify the situation of the system and its environment. The difference is that we focus on the analysis of every single situation and how a task needs to be changed regarding the realization. Dey and Lee collect situations as part of their repertory grid of the problem space. They are focusing on rating features based on the users' preferences in the different situations to identify, which functionalities are flexible and can be changed and which should be available all the time.

The result of *AdaptationExplore* is an adaptation model, which allows the developer to freely choose adaptation mechanisms using a preferred technique or framework.

9   Pilot Study

We piloted *AdaptationExplore* in an exploratory study to explore and initially understand how our newly developed process performs with respect to *effectiveness* and *applicability*. The pilot led also to a couple of insights to outline a follow-up empirical evaluation.

The study was carried out during the Requirements Engineering lecture at the University of Applied Sciences and Arts Dortmund, Germany in the winter term 2019/2020. As a reference for comparison we use a process inspired by Cheng et al. [16] (referred to as *base process* in the following). 37 Masters students from computer science, medical informatics, and business information systems participated in a set of case studies. The students formed groups of 2 to 4 participants, 12 groups in total. The groups were randomly assigned to either *AdaptationExplore* or the base process. Each group was asked to develop a requirements specification of an adaptive system for a problem domain of their own choice. The students were allowed to choose freely in order to ensure sufficient motivation. The scope of the specification work was negotiated with the instructors of the course. The case study took five weeks with an effort of 1.5h/week plus preparation.

With respect to the *effectiveness*, the groups identified between 2 and 4 adaptation points (except for one group G5 that did not find any adaptation point) as Table 2 illustrates. The lack of obvious differences in the quantitative results (between *AdaptationExplore* and the base process and also between the different problem domains) is likely due to a ceiling effect as the identification of adaptation points took place in one of the last weeks of the semester and students were in hurry to prepare deliverables for other courses.

We collected a number of experiences regarding the *applicability*: *Experience E1*: The students liked the tabular goal model identifying high level goals and

Table 2. Number of adaptation points found by groups, G1-G6: *AdaptationExplore*, G7-G12: base process

Group	G1	G2	G3	G4	G5	G6	G7	G8	G9	G10	G11	G12
#Adaptation points	3	3	3	2	0	4	2	2	4	3	3	2

one suggestion was to add priorities to the resources. *E2*: The identification of adaptation points was reported as very easy with *AdaptationExplore*. During the integration of situations, conflicts were identified and marked as adaptation points. *E3*: An issue was the choice of a suitable abstraction level for a situation. One group identified 80+ situations and each contains only a single function On the other side, one group identified three situations which represented the three disjoint functionalities identified during the initial phase. Here the abstraction level was too high.

The pilot study shows that participants need more time to complete the processes, which was difficult to estimate upfront. We gave them 5 weeks, +2 weeks would have been required. An empirical study might need to take place as an *online* study because of the ongoing COVID-19 situation. A study design with a repeated case would be desirable to end up with comparable results for effectiveness and applicability, yet students are used to collaborate intensively online these days.

We plan two empirical studies for a detailed validation of *AdaptationExplore*. The first study focuses on the exploration phase. We want to compare the effectiveness of trigger questions to a general creativity technique used for requirements elicitation in a controlled experiment. The goal is to show the benefits of specific creativity triggers for adaptive systems. In the second study, we compare *AdaptationExplore* with another development process for adaptive systems in a case study to show the advantages of using situations and to further evaluate applicability.

10 Conclusion

This paper introduced *AdaptationExplore*, a novel RE process for adaptive systems with an emphasis on elicitation, negotiation, and documentation. The main idea is to identify situations (defined as system state + environment state) and use a specifically developed set of trigger questions, to identify missing environmental information or alternative realizations to cope with a situation. Other characteristic aspects are the use of goal models along with a new tabular notation as well as the integration of individual instances as a basis for identifying adequate adaptation points.

A pilot study was performed as a exploratory case study with 37 Masters students. We employed the process by Cheng et al. [16] as a baseline for comparison. The results show that *AdaptationExplore* is effective as it helps to identify a considerable number of adaptation points. Based on this study, we also reported on a number of experiences for our process.

For future work, we plan to perform empirical studies to validate the process and its parts in more detail. One study focuses on the exploration phase and the utility of creativity techniques. A second study is planned to compare *AdaptationExplore*, another creativity technique, and the process by Cheng et al. [16].

References

1. de Lemos, R., et al.: Software engineering for self-adaptive systems: a second research roadmap. In: de Lemos, R., Giese, H., Müller, H.A., Shaw, M. (eds.) Software Engineering for Self-Adaptive Systems II. LNCS, vol. 7475, pp. 1–32. Springer, Heidelberg (2013). https://doi.org/10.1007/978-3-642-35813-5_1
2. Berry, D.M.: The tenth anniversary of the CreaRE workshops: a look back and a look forward. In: CEUR Workshop Proceedings, vol. 2584. CEUR-WS.org (2020)
3. Dey, S., Lee, S.-W.: REASSURE: requirements elicitation for adaptive sociotechnical systems using repertory grid. Inf. Softw. Technol. **87**, 160–179 (2017)
4. Qureshi, N.A., Jureta, I.J., Perini, A.: Towards a requirements modeling language for self-adaptive systems. In: Regnell, B., Damian, D. (eds.) REFSQ 2012. LNCS, vol. 7195, pp. 263–279. Springer, Heidelberg (2012). https://doi.org/10.1007/978-3-642-28714-5_24
5. Oriol, M., Qureshi, N.A., Franch, X., Perini, A., Marco, J.: Requirements monitoring for adaptive service-based applications. In: Regnell, B., Damian, D. (eds.) REFSQ 2012. LNCS, vol. 7195, pp. 280–287. Springer, Heidelberg (2012). https://doi.org/10.1007/978-3-642-28714-5_25
6. Sawyer, P., et al.: Requirements-aware systems: a research agenda for RE for self-adaptive systems. In: 18th IEEE International Requirements Engineering Conference (RE), pp. 95–103 (2010)
7. Bencomo, N., et al.: Requirements reflection: requirements as runtime entities. In: 32nd ACM/IEEE International Conference on Software Engineering, vol. 2, pp. 199–202. ACM (2010)
8. Fickas, S., Feather, M.: Requirements monitoring in dynamic environments. In: Second IEEE International Symposium on Requirements Engineering, pp. 140–147 (1995)
9. Carvallo, J.P., Franch, X.: On the use of i* for architecting hybrid systems: a method and an evaluation report. In: The Practice of Enterprise Modeling: Second IFIP WG 8.1 Working Conference, PoEM, pp. 38–53 (2009)
10. Baresi, L., Ghezzi, C.: A journey through SMScom: self-managing situational computing. Comput. Sci. Res. Dev. **28**(4), 267–277 (2013)
11. Bencomo, N., Belaggoun, A.: Supporting decision-making for self-adaptive systems: from goal models to dynamic decision networks. In: Doerr, J., Opdahl, A.L. (eds.) REFSQ 2013. LNCS, vol. 7830, pp. 221–236. Springer, Heidelberg (2013). https://doi.org/10.1007/978-3-642-37422-7_16
12. Alegre, U., et al.: Engineering context-aware systems and applications: a survey. J. Syst. Softw. **117**, 55–83 (2016)
13. Kolos-Mazuryk, L., et al.: A survey of requirements engineeringmethods for pervasive services. In: Freeband A-MUSE Deliverable D5.7 (2006)
14. Paja, E., Maté, A., Woo, C., Mylopoulos, J.: Can goal reasoning techniques be used for strategic decision-making? In: Comyn-Wattiau, I., Tanaka, K., Song, I.-Y., Yamamoto, S., Saeki, M. (eds.) ER 2016. LNCS, vol. 9974, pp. 530–543. Springer, Cham (2016). https://doi.org/10.1007/978-3-319-46397-1_41
15. Soares, M., Vilela, J., Guedes, G., Silva, C., Castro, J.: Core ontology to aid the goal oriented specification for self-adaptive systems. In: New Advances in Information Systems and Technologies. AISC, vol. 444, pp. 609–618. Springer, Cham (2016). https://doi.org/10.1007/978-3-319-31232-3_57

16. Cheng, B.H.C., Sawyer, P., Bencomo, N., Whittle, J.: A goal-based modeling approach to develop requirements of an adaptive system with environmental uncertainty. In: Schürr, A., Selic, B. (eds.) MODELS 2009. LNCS, vol. 5795, pp. 468–483. Springer, Heidelberg (2009). https://doi.org/10.1007/978-3-642-04425-0_36

17. Morandini, M., et al.: Engineering requirements for adaptive systems. Require. Eng. **22**(1), 77–103 (2017)

18. Kneer, F., et al.: Environment modeling for adaptive systems: a systematic literature review. arXiv:2011.07892 (2020)

19. Michalko, M.: Thinkpak. Ten Speed Press (1994)

20. Maiden, N., et al.: Creative requirements: invention and its role in requirements engineering. In: 28th International Conference on Software Engineering, ICSE 2006. Association for Computing Machinery, Shanghai, pp. 1073–1074 (2006)

21. Burnay, C., et al.: Stimulating stakeholders' imagination: new creativity triggers for eliciting novel requirements. In: IEEE 24th International Requirements Engineering Conference (RE), pp. 36–45 (2016)

22. Wieringa, R.J.: Design Science Methodology for Information Systems and Software Engineering, pp. 3–317. Springer, Heidelberg (2014). https://doi.org/10.1007/978-3-662-43839-8

23. Nuseibeh, B., et al.: A framework for expressing the relationships between multiple views in requirements specification. IEEE Trans. Software Eng. **20**(10), 760–773 (1994)

24. Shea, G.: NASA Systems Engineering Handbook Revision 2 (2017)

25. Sutcliffe, A.: Scenario-based requirements engineering. IEEE Trans. Softw. Eng. **24**, 320–329 (2003)

26. Hallsteinsen, S., et al.: Dynamic software product lines. Computer **41**(4), 93–95 (2008)

27. Hinchey, M., et al.: Building dynamic software product lines. IEEE Comput. **10**(45), 22–26 (2012)

Trustworthy AI Services in the Public Sector: What Are Citizens Saying About It?

Karolina Drobotowicz$^{(\boxtimes)}$ ⓘ, Marjo Kauppinen, and Sari Kujala ⓘ

Department of Computer Science, Aalto University, Espoo, Finland
{drobotowicz.karolina,marjo.kauppinen,sari.kujala}@aalto.fi

Abstract. [**Motivation**] Artificial intelligence (AI) creates many opportunities for public institutions, but the unethical use of AI in public services can reduce citizens' trust. [**Question**] The aim of this study was to identify what kind of requirements citizens have for trustworthy AI services in the public sector. The study included 21 interviews and a design workshop of four public AI services. [**Results**] The main finding was that all the participants wanted public AI services to be transparent. This transparency requirement covers a number of questions that trustworthy AI services must answer, such as about their purposes. The participants also asked about the data used in AI services and from what sources the data were collected. They pointed out that AI must provide easy-to-understand explanations. We also distinguished two other important requirements: controlling personal data usage and involving humans in AI services. [**Contribution**] For practitioners, the paper provides a list of questions that trustworthy public AI services should answer. For the research community, it illuminates the transparency requirement of AI systems from the perspective of citizens.

Keywords: Artificial intelligence · Trustworthy AI · Public sector · Transparency · Qualitative research

1 Introduction

Recent advances in artificial intelligence (AI) have popularized this area of research after an "AI winter", a period of waning public interest in AI [1]. AI is also gaining the interest of public organizations due to the opportunities it creates [2], such as reducing administrative burdens and taking on more complex tasks to enable public-sector employees to focus more directly on citizens' needs [3]. The European Commission has also imagined that AI could be used to serve citizens 24/7 in faster, more agile, more accessible ways [4].

However, some public AI services have already harmed society. The AI Now Institute [5] has reported that multiple deployed AI systems have led to misleading results or violations of civil rights. For example, in the United Kingdom,

© Springer Nature Switzerland AG 2021
F. Dalpiaz and P. Spoletini (Eds.): REFSQ 2021, LNCS 12685, pp. 99–115, 2021.
https://doi.org/10.1007/978-3-030-73128-1_7

thousands of immigrants had their visas cancelled due to an erroneous AI system, and in the United States (US) in 2016, AI dramatically lowered the number of home-care hours for people with disabilities without any explanation or possibility to contest its decisions. In 2019, the AI Now Institute published another report [6] documenting cases of automated decision systems used in US public administration. For citizens who were not expecting the use of AI in these cases, they became an unpleasant surprises and decreased their trust [7].

In light of the rise of AI in society and its potentially harmful effects, multiple private and public institutions have published principles and guidelines for ethical AI [8]. However, existing guidelines for ethical AI systems are mostly the results of discussions with industry and academic experts, rarely including citizens' needs and voices.

The goal of this qualitative study is therefore to investigate what kind of requirements citizens have for trustworthy AI services in the public sector. We present findings from 21 interviews with Finnish residents and a design workshop on four public AI services. The data were collected as part of the "Citizen Trust Through AI Transparency" project [9], the goal of which was to provide ethical guidelines for AI usage in the public sector.

The remainder of this paper is structured as follows. First, we review the existing literature on ethical guidelines for AI systems, with a focus on public-service cases. Next, we present our research method and its outcomes. Finally, we discuss the results and limitations of the study and conclude with suggestions for future research.

2   Related Work

Jobin et al. [8] reviewed 84 ethical AI guidelines proposed by industrial and scientific institutions, ten of which targeted the public sector. They found five principles repeated in over half the guidelines: 1) transparency, which aims to increase system explainability, interpretability, or disclosure; 2) justice and fairness, which are connected to mitigating bias and discrimination and enabling challenge or redress; 3) nonmaleficence, which focuses on system security and safety; 4) responsibility, which is often presented alongside accountability and refers to legal liability and integrity; 5) privacy, which mostly relates to data protection and data use and is presented both as a value and a user right.

Across academic guidelines, we found two that focus on interaction with AI systems. First, Amershi et al. [10] presented a set of human–AI interaction guidelines based on documents from industry, scientific AI-design publications, and tests with design practitioners. They suggested how AI systems should behave and what options they should give users during interactions with them. They also mentioned the importance of making systems' functions, performance, reasons, and biases transparent. Second, Rzepka and Berger [11] studied the literature to understand how system and user characteristics influence interactions with systems, finding that transparency positively influences user behavior.

Among guidelines on ethical AI in the public sector, we found two created by research institutes. The Alan Turing Institute [12] presented an extensive set of guidelines in three parts: 1) support, underwrite, and motivate values for a responsible data ecosystem; 2) fairness, accountability, sustainability, and transparency principles for designing and using services; and 3) a process-based governance framework to operationalize these guidelines. The second document came from the Harvard ASH center [3], and explored AI usage in citizen services, suggesting six strategies for the government: 1) make AI part of a citizen-centric program, 2) solicit citizen input, 3) build on existing resources, 4) be data-prepared and tread carefully with privacy, 5) mitigate ethical risks and avoid AI decision making, and 6) focus on augmenting employees, not replacing them.

3 Research Methods

3.1 Overview of the Qualitative Study

We carried out a qualitative exploratory study to answer our research question. We decided to triangulate our data collection because, as suggested by Carter et al. [13], it animated a deeper understanding of the topic and uncovered more detailed answers to our research question. We chose in-depth interviews and a workshop, which are complementary methods according to Kaplowitz and Hoehn [14]. Indeed, during our interviews, the participants felt safer and more focused to share more details, and interactions during the workshop stimulated the participants to share thoughts and needs that did not occur to them during the interviews. The two methods also induced different responses: in the interviews, the participants were more reactive, and in the workshop, they were more creative. Moreover, similarly to the findings of Schlosser et al. [15], the workshop helped uncover broader perspectives on research questions and start topics that are difficult to cover in interviews.

3.2 Study Participants

A total of 21 participants were interviewed (11 women and 10 men). The ages of the participants varied between 18 and 67, with an average of 35. Of the participants, 12 had university degrees, 12 were Finnish, and 9 were immigrants who had stayed in Finland for 3–20 years, with an average of 9 years. When asked for self-estimations of AI knowledge and interest, six of them admitted a poor understanding of AI, eight had a medium level of AI knowledge, three were actively interested in AI, and four were working in the AI field.

Later, eight people participated in the workshop (four women and four men). Six had participated previously in the interviews. Their ages were between 22 and 38, with an average of 28. All had at least bachelor's degrees. Three were born in Finland, and the other five had been in Finland for an average of 6.5 years. One had poor knowledge of AI, three had medium awareness of AI, two were actively interested in AI, and two were working in the field of AI.

3.3   Data Collection

In both the interviews and the design workshop, we used fictional public-service AI examples (Table 1) to help participants understand the scope of possible AI usage and focus their conversations. These sample AI services were generated based on discussions with public-sector representatives in Finland to ensure that they were realistic. The data-collection materials, such as the interview questions and sample AI services, can be seen in the appendix to the master's thesis of the first author [16].

Table 1. Cases used in the data collection.

AI service	Example case	Id
Decision making	AI service that makes a decision about whether the applicant will receive housing	C1
Health prediction	AI service that predicts the mental health problems of citizens and informs a family about it	C2
Impact assessment	Automatic assessment of education impact on pupils, where data collected from children are processed by AI	C3
Fraud detection	AI service used in the social insurance organization to detect financial fraud	C4

Interviews. We conducted 21 interviews between June and July 2019. Each lasted between 45 and 60 min. We designed the interview to be semi-structured, because this provides a good balance between deeply understanding the novel topic and avoiding excessive time consumption, as suggested by DiCicco- Bloom et al. [17]. To assess its validity, the interview process was checked by experienced researchers and piloted with an external person whose answers are not included in the results.

The interviews were divided into three parts. First, the participants were asked general questions about their current attitudes and knowledge about AI in the public sector. Second, they were presented 2–4 public-service AI cases from Table 1. Cases were given in a counterbalanced order to avoid sequence bias. The interviewees were asked to share their concerns, needs, and questions related to each case. All the cases were deliberately information-scarce to nudge the participants to point out what they were missing. The third part of each interview contained a few follow-up questions on using AI in public services.

The participants were invited via physical social security offices and online university channels. We aimed to find people with diverse educations, ages, genders, and AI knowledge. The participants had to be over 18 years old and to have lived in Finland for at least three years. If the person was comfortable speaking English, the interview was conducted in that language; otherwise, it was done in Finnish. Each interview was audio-recorded and transcribed with the consent of the interviewees. A movie ticket was given in exchange for participation.

Design Workshop. The workshop was conducted in July 2019. Its goal was to engage citizens in determining requirements for trustworthy public AI services. The workshop method was inspired by the focus group technique [18] and the ideation methodology described by Michanek and Breiler [19]. The workshop started with a warming-up game. Later, the main task was introduced and repeated three times. Each group found an information sheet by their workstation with a description and example case of AI service in the public sector (Table 1). In the groups, the participants were asked to discuss how to make these services trustworthy. They were asked to save the results of their discussions in whatever form they found useful, using any of blank A3 paper, sticky notes, pens, printed phone mock-ups, and markers. The results of each group's discussion were then visible to the next group coming to the workstation so that the participants could be inspired by previous outcomes. Each round, the participants were put in different groups of two or three people at a new workstation. The workshop ended with a follow-up discussion in which the participants summarized all the results. The workshop was two hours long, and each participant was offered two movie tickets for participating.

3.4 Data Analysis

All the collected data were open and axially coded according to *Research Methods in Human-Computer Interaction* [20]. First, we analyzed the interview and workshop data separately. Next, we compared the results of these two analyses and then combined them. Whenever a result would come solely from AI experts, we recorded it and specified in the text below.

Interviews. Transcribed interviews were analyzed with the support of the qualitative data-analysis software Atlas.ti. The analysis started with one researcher reading the transcripts and marking segments of texts with descriptive in vivo codes. Three types of codes were identified: needs, concerns, and questions. We then iteratively categorized and compared coded segments of data. First, we grouped codes of all forms into high-level concept categories, such as "data" and "purpose." For example, the concept "data" included codes such as "datasource" or "consent." Second, for each category, we read all the included text segments and clustered them into subcategories. For example, in "data," we distinguished subcategories such as "data collection" and "data bias."

Design Workshop. Data from the workshop were saved in the forms of physical sticky notes and an audio recording of the follow-up discussion. The analysis of the workshop materials was similar to the interview analysis but did not employ any digital tools. First, we reviewed sticky notes while listening to the audio recording to clarify and add missing information. Next, we clustered sticky notes by the high-level concepts to which they were related. We then examined the notes inside each cluster and divided them into subcategories. Like the interview analysis, this was also done iteratively.

Axial Coding. We compared the subcategories of the interview and workshop analyses and identified similarities, differences, and relationships between them. As suggested by Charmaz [21], this iterative process enabled a deeper understanding of the concepts and thus improved the accuracy of our results.

4   Results

In this section, we present the requirements that our participants shared for trustworthy public AI services. A significant part of the discussions with the participants was focused on transparency, so we start by introducing this requirement. We continue by presenting the participants' detailed questions and requirements grouped into five concepts: purpose, data, core AI process, human involvement, and service overview.

4.1   Transparency

All the participants wanted to know more about the public AI services than was presented in the materials. Regarding motivation, they referred to uncomfortable emotions (e.g., "I fear AI if it collects something that is not said. Transparency throughout the research process is needed, otherwise it can feel bad."), the need to make informed decisions (e.g., "If I'm convinced how they get the results, [...] then I can decide"), and trust (e.g., "They don't have to give me all the information at all times, but [they should] be transparent on how they process the information so that I have more trust in them.").

4.2   Purpose

The participants asked multiple questions related to the purposes of services (Table 2). First, many participants highlighted their need to know the topical purposes of public AI services presented in the study. In the follow-up discussion, one interviewee explained, "So, there should be transparency about purpose. What is the intended purpose? What is the base reason this service exists?" Knowledge about the purposes of the services was especially required in the impact-assessment case (C3), where questions like the following emerged: "What are the targets of the project?"

Table 2. Three questions describing purpose-related requirements.

Subcategory	Questions
Purpose	*For what* reason was the public service created?
Benefits	*What* are the benefits that the public service brings?
Impact	*What* impact on users or on society can the public service make?

Several participants asked more specifically about the potential benefits of the public services for them and other stakeholders: "What I am expected to benefit from this information?" (C2) and "Would the children benefit from this? Or parents? What is the benefit of the school?" (C3). A few participants also stated that if a service presented a clear benefit to them, they would be more inclined to use it, even if it was not fully transparent. For example, in the health-prediction case (C2), one interviewee said, "A grandmother's well-being is more important than where the data come from."

Lastly, a few inquiries were made about the impacts of the public services in the education-related case (C3). Participants asked, "What is the social impact for the participants?", "If you are part of the experiment, you would like to know what it is for you in the future. How much does it affect your position in society?" Two participants working actively with AI mentioned that public AI services should increase rather than decrease social justice. They mentioned examples of AI methods, such as scoring and grouping, that should not happen in the public sector; for example, "If a child is from a different background and gets results which seem bad, then they might be put in some special group for slow people. But it could just be a misunderstanding of questions or [a] different background. Then, you're limiting that child's abilities to do well in the future."

4.3 Data

Data collection is an essential part of any AI service, and the study participants had multiple questions about it. We categorized these questions into six subcategories (Table 3). Notably, the participants focused their questions on personal data due to the specifics of the presented cases.

Table 3. Eleven questions describing data-related requirements.

Subcategory	Questions
Data source	*What* is the source of data collected in the public AI service?
Data collection	*When* and *how* were the data about the user collected? *When* was the consent given for collecting this data?
Data purpose	*Why* was this specific information needed?
Data storage	*Where* and *for how long* are the data stored?
Data access	*Who* has the access to the data?
Data bias	*Are* the data biased? *Why*? *How* do they impact the results?

First, many interviewees started by asking about the sources of the data in the public AI services. This was especially relevant to the case of automatically prefilled applications for housing (C1) and fraud discovery (C4). For example, two interviewees mentioned, "I'd like to find out where the information came from. It's irritating when not told here," and, "Where do they have the data from?" Because there was no information on the data sources during the interviews, participants shared their own guesses and attitudes. In the first case (C1), participants were mostly sure that it came from other public organizations. They shared a positive attitude about that because, in their opinions, it could make the process easier and faster: "I did it all online. And they brought all of the data [from other institutions]. This is really cool because it saves me a lot of time [...]. I knew exactly where they were getting the data from. So, didn't bother me." However, the fourth case (C4) incited more controversy, as participants guessed that private companies were the data sources. That case provoked more opposing voices, such as, "Maybe they get my income and spending from my bank? I don't think they should do that because that's crossing the border from the public to the private sector." These outnumbered the accepting voices, such as, "It doesn't hurt even if the information is borrowed from somewhere else."

Several participants also wanted to know more details about the data collection. A few interviewees asked about whether and when they gave their consent to share their data: "Where and when do I consent to this? If I didn't consent, then why are they collecting?" The workshop attendants were also interested in how and from what period data were actually collected. According to them, it was also essential to know why specific data are collected for public services. This was especially relevant in the impact-assessment case (C3), in which data were collected from children. One interviewee asked, "What is the justification behind collecting this much information on my and other [people's] children?"

When data are already collected, they must be stored somewhere; the participants with greater AI knowledge were interested in this topic as well. They asked where and for how long they were stored: "How and where are the data stored, and in what kind of format?" The workshop attendants added that they wanted to know what happened to user data after the services were finished and what organizations or people had access to their data: "[I should] know where the information was going." A few interviewees also asked how they could access their collected data.

Lastly, there was some discussion about data bias, that is, how using unrepresentative data can lead to discriminating results. This topic was rarely started by interviewees, possibly due to their lack of knowledge in this area. During the workshop, three of the eight participants who worked with AI or were interested in it knew what AI bias was. Upon discussion, participants suggested the importance of informing users about possible biases in AI systems, why they emerge, and how they can affect results.

Apart from their questions, many participants stated their requirements related to data. First, they highlighted the importance of consent to share data: "[I want to] decide whether or not I consent to some information being collected on me." Next, some suggested that after data are already collected, they should be able to review them. According to interviewees, this would enable users to notice any problems with the data, such as being too old, missing something, or being wrong: "If they collect the data, you should have some sort of report. You could say when something is missing." A few participants also requested full control over their data: "It should be possible to keep track of where the information goes. So, even though it would mean that I won't be favored in certain decisions, I'd still like to control information that is given." "We should be more aware of what our data [are] being used for. And we should be in more control of switching on and switching off what we do and do not share."

Participants also shared concerns about their privacy and the security of their data. They especially opposed too much information being collected about their children and relatives in cases C2 and C3 and their financial status in C4: "[It] wouldn't be okay to see messages they send to the family example," "knowing that your grandma might be in danger of social exclusion just sounds like there's a constant surveillance on her," and "it feels like a privacy violation." They also shared the requirement of storing personal data securely: "Security is really important in a lot of these. Because [...] it can also be exploited by companies to do targeting. Or it can be exploited by the government." "There should be an assurance that no one can access your information." One interviewee with extensive AI knowledge shared another concern: "The only thing that worries me in the public sector is that: Do we have the best people to keep the data protected?".

4.4   Core AI Process

We distinguished the core AI process inside each public service responsible for creating its results, that is, the intentional output generated by each service, such as a decision or prediction. In this section, we describe the subcategories that we grouped around the concept of the core AI process (Table 4).

Table 4. Seven questions describing core AI process related requirements.

Subcategory	Questions
AI process reason	*What* is the reason for using AI in the public service?
Used criteria	*What* criteria are being used for the results creation?
Used data	*What* data are used for results creation?
Results creation process	*What* is the process of results creation?
Results explanation	*What* is the reason for the results? *Which* data and criteria affected the results?
Results reliability	*How* reliable are the results?

First, the participants requested to know the reason for using each AI process, especially in the impact-assessment case (C3): "Why this way? What is the justification behind collecting this much information on my children and other children? And why does it have to be [this way]?" It was also of interest to the participants to know what criteria were used to create the results: "I would certainly be very interested in what the parameters are that affect the decision" and "What kind of laws [do] they have for [the] particular benefit that I'm applying for?" Next, many participants asked what data were actually used to create the results: "What information would be utilized?" and "It would be useful to know what information is used." They also asked much more detailed questions about the data, which are presented above (Sect. 4.3).

Most of the participants asked about the AI process, that is, what is actually done with the criteria and data to create the results. For example, interviewees asked, "How do they do [the process]? How did they use your data?" "What kind of conclusions are they trying to get out of it?" and "[It would be] good to know how they analyzed this case." According to the participants, it was vital that they at least know that AI is used in the process: "I think there's no reason to hide that [AI is used] because I think some people certainly would have negative feelings if they didn't know" and "I feel awkward; I was tricked. [...] If I know that it's an automated process, I will feel better."

The participants also mentioned the need to control the process. During the workshop, participants suggested that each service should have options, such as always being able to quit the service or to have humans handling tasks instead of an AI. Interviewees also shared their worries over not controlling services and AI in general: "I don't think I can control [what happens in the service]" and "I don't think I can stop or make [AI] more humane; it's going too fast."

When participants were given the results of the AI services presented in the study, they often asked for explanations, especially in the decision-making case (C1). Two responses were, "What are the exact reasons?" and "There's a lack of information as to why my application is rejected." A few participants suggested how the explanation should look: "It would need to be professional and have clear indications what the exact reasons are and reference to certain clauses" (C4), "Something like, we have a list of people that have been waiting for a long time, or the refugees, some reason you can understand" (C1), and "Why is it rejected? Like, [...] there is no free apartment. My wishes are too big" (C1). Lastly, one participant commented on the reason for an explanation: "[I would] have to call somebody to try to figure out why. Then they also have to try to figure out why. So, if it is smart enough to decide immediately why I'm not going to get the house, it should also be smart enough to tell me immediately why."

Participants more experienced in AI also requested knowledge about result reliability, that is, their accuracy and trustworthiness. One asked, "How confident are the results? Like, are they 110% confident? Or is it more like it might be that the system works, or it's like 100% prediction, or in a hundred thousand cases before me, this happened?" During the workshop, participants also suggested that especially in the healthcare services it should be clearly written how much the results could be trusted, such as by stating, "This is not a diagnosis and does not replace medical professionals." They affirmed that it is vital to provide levels of confidence in results, as they might be erroneous.

4.5 Human Involvement

Human personnel are usually involved in public-service operations, but they can also be involved in core AI processes. The need to know where actual humans are involved in creating results was highlighted in the workshop, although few interviewees asked for it (Table 5). However, interviewees shared multiple needs and concerns related to the roles of humans in AI services.

Table 5. One question describing human-involvement related requirements.

Subcategory	Questions
Human involvement	*What* is the role of humans in the results creation process?

First, the interviewees shared the general need to interact with people instead of AI, especially in personal cases, such as healthcare. One interviewee said, "We can replace as many things as possible with machines. But at the end of the day, we still crave human interaction in some form or another." According to the participants, human personnel could be responsible for introducing people to the service or explaining its results: "If someone is telling face-to-face, it's easier to motivate or convince the person. But if it's some odd papers, sometimes you just skip the part that you didn't need." During the workshop, it was also mentioned that there should be an easy way to contact someone from the service.

Second, the interviewees were concerned that AI would not be able to understand a human case as well as another human, as it overgeneralizes and lacks human intuition: "Especially in healthcare, I want humans to talk to because there are a lot of things that are not possible to be read by the program." In application forms, as in C1, one person suggested such a solution: "I would honestly prefer that there was an open field to describe your life situation right now. And then there would be a human in the loop looking at the application."

A few participants suggested that it would be better to have humans make final decisions, especially when they are important. Such comments were given in the follow-up discussion and on the fraud-detection case: "It is worrying if solely AI would be taking decisions on humans' lives" and "I would like a person to see and decide based on this information rather than artificial intelligence" (C4). One person highly educated in AI added, "As of now, we are still stumbling upon training AI to the point that it does the decisions correctly [...]. I still don't feel comfortable [with an AI] making the decision on its own."

Lastly, several participants suggested having human controllers monitor AI services for possible errors, rule-breaking, and unethical actions. One interviewee who worked with AI said, "Nothing should be [fully] automated. When it comes to analysis and evaluation, you have to have someone who can verify that the system is working according to rules and ethical guidelines, as demanded by society." The workshop participants suggested always having the option to ask for a human review of an AI process.

4.6   Service Overview

The service overview contains general, practical knowledge about each service that participants asked about (Table 6). First, the study participants mentioned their interest in understanding the high-level processes of the public services. Some interviewees asked general questions: "How does this service work in practice?" Others asked more case-specific questions: "How frequently and how will it be available?" During the workshop, participants requested basic information about the service stages and how long they take. They also requested updates on service statuses when results were not immediate, such as in C3. Two interviewees said, "Maybe every half a year, or maybe even once a month" and "It would be good every six months to get follow-up information."

Table 6. Five questions describing high-level service related requirements.

Subcategory	Questions
High-level process	*What* should customers expect from the public service?
	What are the stages of the service and *how long* do they take?
Accountability	*Who* is accountable for the public service?
Users of the service	*Who* are the users of the public service?

Some interviewees were also interested in knowing the other users of the service. For example, two interviewees asked about the number of other children whose data would be collected for the educational impact-assessment case (C3): "Is it only my child? Is it the whole class?" and "I would like to know how many other children are involved." Lastly, participants asked for who or what organization was accountable for the service and its outcomes: "Who has decided?", "Who has developed it?" and "What is this social welfare organization?".

5   Discussion

5.1   Transparency

The most important finding of this study is that transparency is a critical requirement for trustworthy AI services from the perspective of citizens. This result is consistent with the finding of Jobin et al. [8], who report that transparency is the most common principle across ethical AI guidelines. However, it can be a demanding task to specify the transparency requirement systematically in practice. For example, multiple transparency definitions have been proposed in the AI services context. In this study, we focused on the visibility of the service information and justifiability of AI service processes and outcomes, as defined by Leslie [12] and Turilli et al. [22]. In more detail, Turilli et al. [22] suggested that transparency should explain the processes accomplished by the service (how, by whom, and what was collected and done), as it enables checking whether the service is a product of ethical processes. Hosseini et al. [23] suggested that to reach meaningful transparency, services must be open about policy (why), process (how), and data (what). Our study contributes to these by providing 27 detailed questions that should be answered by AI services in the public sector for citizen trust. We discuss those questions below.

First, the participants were interested in the purposes of the services, why they existed and what impacts they had on them and others. This was especially important when their benefits were not clear. Second, the participants asked multiple questions about data: from what sources and how the data were collected and whether data owners consented to give the data. They also had privacy-related questions, such as who could access their personal data and how they would be stored. Only a few participants raised the topics of bias and fairness, even though it was one of the most common principles found by Jobin et al. [8], perhaps indicating that those topics are not well known among non-specialists.

Third, the participants shared multiple questions about core AI processes. They were interested in what data and criteria were used and how they were processed to create results. This information was relevant for participants both before they joined a service and as an explanation of its results. This supports the findings of Chazette et al. [24], who found that the vast majority of their survey respondents found service-result explanations necessary. Furthermore, they found that "what" and "why" questions were more important in explanations than "how." A few participants in our study also specified that explanations must be easily understandable by non-specialists, a requirement pointed out in the public-sector guidelines from Alan Turing Institute [12]: explanations should be socially meaningful and devoid of technical language. Fourth, the participants asked questions about the roles of humans in creating results, and fifth, they asked about service overviews.

The results of this study show that AI transparency is very closely related to AI explainability, which has been studied extensively. For example, Arrieta et al. [25] performed the literature review of approximately 400 publications related to explainable AI and defined explainability as "the details and reasons a model gives to make its functioning clear or easy to understand." They also presented explainable AI as a core element needed to achieve responsible AI principles, including transparency. Similarly, Chazette et al. [26] discovered that explainability was the means to achieving the non-functional requirement of transparency. Our study revealed the detailed citizen requirements for explainable AI, such as the visibility of the criteria and data used by the AI and the understandable explanation of results produced by AI.

5.2 Other Requirements

Apart from transparency, the participants shared other requirements. We discuss the two most important here. First, they highlighted the need to have humans involved in services, although participants' views on this diverged. Some emphasized being able to interact with a person to discuss a service, while others only wanted people to be involved in reviewing their data and making decisions or in monitoring the whole AI process. Second, the participants required a certain level of control over their data. Most often, they wanted to be asked for consent before any of their data were collected or shared. A few also requested full control over their data, to be able to choose which data are used, and to be able to withdraw them at any point. Part of these requirements are mentioned in the Harvard ASH Center's strategies for government and public institutions [3], which state that asking citizens for consent to use their data in services creates fewer privacy concerns, discourage letting only AI make critical decisions for citizens, and encourage human oversight.

5.3   Study Limitations

Generalizability. We interviewed only residents of the Metropolitan Area of Finland between the ages of 18 and 67. Despite our efforts to include diverse participants, we cannot confirm that other demographic groups would have similar requirements. In fact, in different parts of the globe, societies have different cultural biases and values that influence their mental models of AI [27]. Even within Europe, cultural and social characteristics vary [28]. Finland, for instance, enjoys greater trust in the public sector [29].

Reliability. We are also aware of the inherent weaknesses of the interview and workshop techniques. For one, the interviewers may have passed their occupational biases into the research. Interviewees also may not have told the truth or not understood the questions well. However, we took a few precautions to counter these threats to reliability. First, our interview questions were reviewed by senior researchers and were piloted. Second, the data-analysis process was reviewed by another senior researcher. Third, the participants came voluntarily for the interviews, they did not need to answer every question, and they were informed that what they said would remain confidential and anonymous.

Six participants took part in both the interviews and the workshop, and we are aware of the bias they may have brought to the workshop by changing or emphasizing opinions they stated during their interviews. However, because of diverse interactions during the workshop, these six participants had chances to discuss topics not covered during the interviews. Moreover, we believe this diversity in the topic awareness likely positively affected results of a workshop by inducing more perspectives to the discussions.

Lastly, we included four AI specialists and three people actively interested in AI in our study. To reason it, we need to share the Finnish AI context. In 2018, Finland released the estimation that one-fifth of its population would eventually need to obtain AI skills [30]. By now, more than 1% of Finnish citizens have expanded their knowledge of AI by taking the freely available course "Elements of AI," and Finnish universities altogether offer 250 AI courses, which are taken by about 6,300 students every year [31]. Finally, we believe those AI specialist are citizens, whose voices are also valuable and who may actively shape future AI in the public sector. For clarity, we also marked all the results that came from only this group of participants.

6   Conclusions

This paper presents citizens' requirements for trustworthy AI services in the public sector. Based on our findings, transparency is a particularly important requirement of public AI services. Specifically, for practitioners, this paper provides a list of 27 questions that ought to be answered by such services to achieve trustworthiness. The results of this study also indicate that citizens have other important requirements, such as the need to control one's data and to have humans involved in AI processes. We suggest that these questions and requirements guide public AI service design and development. For the research community, we contribute by extending the knowledge of the transparency requirement of AI systems from the perspective of citizens.

Reflecting on our experience, we suggest the following for future research. The findings of this paper could be tested with citizens in the form of public AI service prototypes to validate our results and study the depth of information required by citizens to optimize transparency. As another direction, the study of citizen requirements could be broadened by including the private sector. For example, the healthcare sector may be an interesting area to study, as it includes both private and public organizations and is proximal to citizens.

Acknowledgements. We thank the Saidot team from spring 2019 for starting the project and assisting in the data collection of this study, J. Mattila for co-organizing and conducting parts of the interviews, and our participants for sharing their experiences.

References

1. Fast, E., Horvitz, E.: Long-term trends in the public perception of artificial intelligence. In: Proceedings of the Thirty-First AAAI Conference on Artificial Intelligence, pp. 963–969 (2010)
2. Cath, C., Wachter, S., Mittelstadt, B., Taddeo, M., Floridi, L.: Artificial intelligence and the 'Good Society': the US, EU, and UK approach. Sci. Eng. Ethics **24**, 505–528 (2017)
3. Mehr, H.: Artificial Intelligence for Citizen Services and Government. Harvard Ash Center Technology & Democracy (2017)
4. AI HLEG, Policy and investment recommendations for trustworthy AI. European Commission (2019)
5. AI Now Institute: AI Now Report 2018 (2018)
6. AI Now Institute: Automated Decision Systems Examples of Government Use Cases (2019)
7. New York City's algorithm task force is fracturing. https://www.theverge.com/2019/4/15/18309437/new-york-city-accountability-task-force-law-algorithm-transparency-automation. Accessed 6 Nov 2020
8. Jobin, A., Ienca, M., Vayena, E.: The global landscape of AI ethics guidelines. Nat. Mach. Intell. **1**(2), 389–399 (2019)
9. A Consortium of Finnish organisations seeks for a shared way to proactively inform citizens on AI use. https://www.espoo.fi/en-US/A_Consortium_of_Finnish_organisations_se(167195). Accessed 6 Nov 2020

10. Amershi, S., et al.: Guidelines for human-AI interaction. In: Proceedings of the 2019 CHI Conference on Human Factors in Computing Systems, pp. 1–13 (2019)

11. Rzepka, C., Berger, B.: User interaction with AI-enabled systems: a Systematic review of IS research. In: Proceedings of the 39th International Conference on Information Systems, pp. 13–16 (2018)

12. Leslie, D.: Understanding artificial intelligence ethics and safety: a guide for the responsible design and implementation of AI systems in the public sector. Alan Turing Institute (2019)

13. Carter, N., Bryant-Lukosius, D., Dicenso, A., Blythe, J., Neville, A.: The use of triangulation in qualitative research. Oncol. Nurs. Forum **41**(5), 545–547 (2014)

14. Kaplowitz, M., Hoehn, J.: Do focus groups and individual interviews reveal the same information for natural resource valuation? Ecol. Econ. **36**(2), 237–247 (2001)

15. Schlosser, C., Jones, S., Maiden, N.: Using a creativity workshop to generate requirements for an event database application. In: Paech, B., Rolland, C. (eds.) REFSQ 2008. LNCS, vol. 5025, pp. 109–122. Springer, Heidelberg (2008). https://doi.org/10.1007/978-3-540-69062-7_10

16. Drobotowicz, K: Guidelines for designing trustworthy AI services in the public sector. Master's thesis, Aalto University, Department of Computer Science (2020). http://urn.fi/URN:NBN:fi:aalto-202008235015

17. DiCicco-Bloom, B., Crabtree, B.: The qualitative research interview. Med. Educ. **4**(4), 314–321 (2006)

18. Kitzinger, J.: Qualitative research: introducing focus groups. BMJ **311**(7000), 299–302 (1995)

19. Michanek, J., Breiler, A.: The Idea Agent: The Handbook on Creative Processes, 2nd edn. Routledge, Abingdon (2013)

20. Lazar, J., Feng, J., Hochheiser, H.: Research Methods in Human-Computer Interaction, 2nd edn. Morgan Kaufmann, Burlington (2017)

21. Charmaz, K., Hochheiser, H.: Constructing Grounded Theory: A Practical Guide Through Qualitative Analysis, Thousand Oaks (2006)

22. Turilli, M., Floridi, L.: The ethics of information transparency. Ethics Inf. Technol. **11**(2), 105–112 (2009). https://doi.org/10.1007/s10676-009-9187-9

23. Hosseini, M., Shahri, A., Phalp, K., Ali, R.: Foundations for transparency requirements engineering. In: Daneva, M., Pastor, O. (eds.) REFSQ 2016. LNCS, vol. 9619, pp. 225–231. Springer, Cham (2016). https://doi.org/10.1007/978-3-319-30282-9_15

24. Chazette, L., Karras, O., Schneider, K.: Do End-Users want explanations? Analyzing the role of explainability as an emerging aspect of non-functional requirements. In: Proceedings of the IEEE International Conference on Requirements Engineering, pp. 223–233 (2019)

25. Arrieta, A.B., et al.: Explainable artificial intelligence (XAI): Concepts, taxonomies, opportunities and challenges toward responsible AI. Inf. Fusion **58**, 82–115 (2020)

26. Chazette, L., Schneider, K.: Explainability as a non-functional requirement: challenges and recommendations. Requirements Eng. **25**(4), 493–514 (2020). https://doi.org/10.1007/s00766-020-00333-1

27. Schaefer, K., Chen, J., Szalma, J., Hancock, P.: A meta-analysis of factors influencing the development of trust in automation: implications for understanding autonomy in future systems. Hum. Factors **58**(3), 377–400 (2016)

28. Lee, J., See, K.: Trust in automation: designing for appropriate reliance. Hum. Factors **46**(1), 50–80 (2004)

29. Leading the way into the age of artificial intelligence Final report of Finland's Artificial Intelligence Programme 2019. Publications of the Ministry of Economic Affairs and Employment (2019)
30. Koski, O.: Work in the age of artificial intelligence: Four perspectives on the economy, employment, skills and ethics. Publications of the Ministry of Economic Affairs and Employment of Finland (2018)
31. Artificial Intelligence From Finland, e-Book of Business Finland (2020). https://www.magnetcloud1.eu/b/businessfinland/AI_From_Finland_eBook/

Defining Utility Functions
for Multi-stakeholder Self-adaptive
Systems

Rebekka Wohlrab$^{(\boxtimes)}$ and David Garlan

School of Computer Science, Carnegie Mellon University, Pittsburgh, USA
wohlrab@cmu.edu, garlan@cs.cmu.edu

Abstract. [**Context and motivation:**] For realistic self-adaptive systems, multiple quality attributes need to be considered and traded off against each other. These quality attributes are commonly encoded in a *utility function*, for instance, a weighted sum of relevant objectives. [**Question/problem:**] The research agenda for requirements engineering for self-adaptive systems has raised the need for decision-making techniques that consider the trade-offs and priorities of multiple objectives. Human stakeholders need to be engaged in the decision-making process so that the relative importance of each objective can be correctly elicited. [**Principal ideas/results:**] This research preview paper presents a method that supports multiple stakeholders in prioritizing relevant quality attributes, negotiating priorities to reach an agreement, and giving input to define utility functions for self-adaptive systems. [**Contribution:**] The proposed method constitutes a lightweight solution for utility function definition. It can be applied by practitioners and researchers who aim to develop self-adaptive systems that meet stakeholders' requirements. We present details of our plan to study the application of our method using a case study.

Keywords: Self-adaptive systems · Quality attributes · Utility functions · Analytic Hierarchy Process

1 Introduction

For self-adaptive systems, multiple quality attributes (such as performance, availability, and security) need to be considered and traded off against each other. These quality attributes are often encoded in a *utility function*, i.e., a single aggregate function whose expected value should be maximized by the system [4,6,8,16]. In self-adaptive systems, utility functions are typically used by automated planning mechanisms to identify the relative costs and benefits of alternative strategies. In related work, utility functions are often defined as the weighted sum of relevant objectives [3,5,18]. For most approaches using utility functions, it is simply stated that they should be manually defined, but little guidance for this task is provided [8,18]. It is challenging to correctly identify

© Springer Nature Switzerland AG 2021
F. Dalpiaz and P. Spoletini (Eds.): REFSQ 2021, LNCS 12685, pp. 116–122, 2021.
https://doi.org/10.1007/978-3-030-73128-1_8

the weights of each objective and consider trade-offs between multiple quality attributes, as reported in the research agenda for requirements engineering for self-adaptive systems [16]. Self-adaptive systems often have multiple stakeholders (e.g., end users or business owners) whose preferences need to be consolidated to identify the overall relative importance of each objective [15]. Decision-making techniques are needed to help stakeholders prioritize and negotiate quality attributes and determine appropriate utility function weights [16].

In this paper, we present a lightweight tool-supported method for utility function definition for multi-stakeholder self-adaptive systems. The proposed method is based on the Analytic Hierarchy Process (AHP) [14] and the Delphi technique [7]. It supports stakeholders in prioritizing quality attributes, negotiating priorities to reach an agreement, recording rationales and comments, and giving input to define utility functions. We use the weighted sum approach for utility functions, as it is lightweight and commonly used in related work [3,5,18]. It assumes that the weighted quantity of one quality attribute can be traded off (or "substituted" [1]) with another one. Guidelines to select utility functions, considering risk and dependencies between decision parameters, have been previously created [1] and can be used to support other kinds of utility functions in the future. The proposed ideas of this research preview paper will be further refined and we plan to study the method's application in a case study. We expect that our method will be of use to practitioners and researchers that aim to conceive self-adaptive systems fulfilling stakeholders' preferences.

2   Proposed Approach

Figure 1 shows the steps of our method for utility function definition. The method can either be used for the initial definition or the refinement of the utility function, in case stakeholders' preferences evolve over time. The two leftmost steps are performed individually by each stakeholder. The two guard conditions refer to whether an AHP matrix is inconsistent and whether no agreement has been reached. Each step is labeled with the paragraph in which it is described.

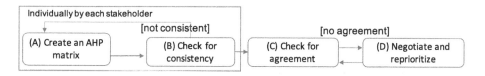

Fig. 1. Overview of our method for utility function definition

(A) Create an AHP Matrix: For the prioritization of quality attributes, we use the AHP, which is especially useful when subjective, abstract, or non-quantifiable criteria are relevant for a decision [14]. A central part of the AHP is to elicit stakeholders' priorities of different objectives in pairwise comparison matrices, which are positive and reciprocal (i.e., $a_{ij} = 1/a_{ji}$). For utility functions, we are interested in the degree of preference of one quality attribute over another, with the goal of increasing the overall utility of a system. Verbal expressions are used for these pairwise comparisons (e.g., "*I strongly prefer X over Y*"). Table 1 shows how the verbal expressions correspond to numerical values.

For a robot planning problem, Table 2 shows an example of an AHP matrix with the attributes safety (expected number of collisions), speed (duration of a mission), and energy consumption (consumed watt-hours). In the example, safety is *very strongly* preferred over speed (7) and *extremely preferred* over energy consumption (9). Speed and energy consumption are *equally preferred*.

The relative priorities of the quality attributes can then be calculated using the principal eigenvector of the eigenvalue problem $Aw = \lambda_{max} w$ [14]. A is the matrix of judgments and λ_{max} is the principal eigenvalue. For the matrix in Table 2, the principal eigenvalue is $\lambda_{max} \approx 3.01$. A corresponding normalized eigenvector to λ_{max} is $(0.8, 0.1, 0.1)^T$, which corresponds to the relative priorities of the quality attributes. The utility function for a mission might be defined as $U(m) = 0.8 \cdot \mathsf{safety}(m) + 0.1 \cdot \mathsf{duration}(m) + 0.1 \cdot \mathsf{energy}(m)$. $\mathsf{safety}(m)$ indicates the expected number of collisions in a mission, $\mathsf{duration}(m)$ the number of timesteps, and $\mathsf{energy}(m)$ the consumed watt-hours. The preference of a quality attribute can often be described with a sigmoid function defining an interval for the quantity that is considered as good enough and an interval for the quantity that is insufficient [12]. Appropriate methods will need to be created in the future to elicit these thresholds and define quality attributes' preference functions.

Table 1. AHP judgment/preference options with numerical values [14].

Extremely preferred	9
Very strongly preferred	7
Strongly preferred	5
Moderately preferred	3
Equally preferred	1
Intermediate values	2, 4, 6, 8

Table 2. Example of an AHP matrix.

	Safety	Speed	Energy Consumption
Safety	1	7	9
Speed	$\frac{1}{7}$	1	1
Energy Cons	$\frac{1}{9}$	1	1

(B) Check for Consistency: AHP matrices can be checked for consistency. A matrix is consistent if $a_{jk} = a_{ik}/a_{ij}$ for $i, j, k = 1, \ldots, n$ [14]. Saaty proved that a necessary and sufficient condition for consistency is that the principal eigenvalue of A be equal to n, the order of A [14]. He defined the consistency index CI as $(\lambda_{max} - n)/(n - 1)$. For our example in Sect. 2, CI is 0.004. To compare consistency values, Saaty also calculated the *random consistency index* RI by calculating CI for a large number of reciprocal matrices with random entries [14].

For a 3×3 matrix, the average random consistency index was 0.58. According to Saaty, the consistency ratio $CR = CI/RI$ shall be less or equal to 0.10 for the matrix to be considered consistent [14]. In our example, the consistency ratio is 0.01. If consistency is not fulfilled, stakeholders are required to refine their AHP matrices. The matrix can be automatically analyzed to point out the triples of quality attributes QA_i, QA_j, and QA_k where $a_{jk} \ll a_{ik}/a_{ij}$ or $a_{jk} \gg a_{ik}/a_{ij}$.

(C) Check for Agreement: We consider the rankings of n quality attributes by k stakeholders (where each quality attribute's rank is a number between 1 and n). For QA_i, the sum of ranks by all stakeholders is R_i, and the mean value of these ranks is $\bar{R} = \frac{1}{n} \sum_{i=1}^{n} R_i$. If the stakeholders' rankings do not agree, we can assume that the sums of ranks of several quality attributes are approximately equal [10]. It is therefore natural to consider the sum of squared deviations from the mean values of ranks $S = \sum_{i=1}^{n} (R_i - \bar{R})^2$ [10]. The maximum possible value of S is $k^2(n^3 - n)/12$ [10]. Kendall's concordance coefficient, describing the agreement of rankings in a [0,1] interval, is therefore: $W = \frac{12S}{k^2 \cdot (n^3 - n)}$ [10].

(D) Negotiate and Reprioritize: In case agreement is not reached, a tool-supported negotiation and reprioritization phase starts. To aggregate AHP matrices, the "most recommendable aggregation technique" is to calculate the weighted arithmetic mean of individual priorities (AIP) [11]. A priority indicates the importance of a quality attribute with a value between 0 and 1. Stakeholders' priorities can be weighted differently, as their influence and stake may differ.

While the AIP can be used to quickly arrive at a solution, it is beneficial to discuss and record underlying rationales. We adapt the *Delphi technique* [7] for remote consensus building. Interactive tooling is used to support the technique and collect data. The stakeholders anonymously provide input in several iterations and receive controlled feedback. Users can declare that they do not know or do not care about a quality attribute. It is also possible to delegate votes to another participant (proxy voting). In the first round, open-ended questions are used concerning participants' rationales (e.g., "in what situation(s) do you think safety is especially important?"). The answer is fed back to other participants to inform their rankings. The main trade-offs and conflicts between quality attributes are elicited and discussed. While we assume an existing set of quality attributes, participants can also suggest new quality attributes and objectives.

In the second round, the comments and rationales are presented to the participants and the AHP matrices can be revised. The rankings that are in conflict are indicated to increase transparency. Further comments and rationales are added and a consensus starts to form. In the third round, participants are asked to revise their judgments or declare why they decide to remain outside the consensus [7]. The final utility function is a weighted sum of the objectives, where the final weights are the participants' aggregated weighted priorities (using AIP).

3   Empirical Study

We plan to perform a multiple case study [13] focusing on the phenomenon of applying our approach in practice. The approach depends on contextual factors and is therefore difficult to study in controlled settings (e.g., experiments). As mentioned before, utility functions are a central mechanism in several approaches for self-adaptive systems. We plan to apply the method to existing systems and projects. As the first case, we focus on robot mission planning using a probabilistic model checker, where the correct definition of the weights of multiple objectives is essential. The participants operate from multiple locations and are aware of the system's context and a preliminary set of quality attributes. The stakeholders have conflicting objectives (e.g., end user, business/cost, performance, and safety concerns) and are asked to apply our method for utility definition. We aim to use collected tool data, observations, and complementary interviews to study the required time to build a consensus, the understandability of the approach, as well as negotiation strategies. The empirical study is intended to give insights into how participants typically act to reach a consensus, how beneficial our proposed method is perceived for utility function definition, and how satisfied stakeholders are with the resulting utility function.

4   Related Work

Identifying and prioritizing objectives for self-adaptive systems is a nontrivial task. The Goal-Action-Attribute Model requires the goals, priorities, and preferences of multiple stakeholders to be elicited [15]. The AHP is suggested to be used for this task, but no concrete guidance is given. We focus on prioritization and negotiation and develop a comprehensive AHP-based method. Rather than focusing on creating complete goal models, we aim to create a lightweight method for utility function definition. For security requirements, the Swing-Weight Method has been used for prioritization and utility function definition [2]. The AHP allows a more precise elicitation of the relative priorities of objectives.

Utility functions are a common mechanism in self-adaptive systems [3–6,18]. A few approaches for utility function definition are related to our work on the prioritization and negotiation of utility function weights. Song et al. [17] propose to collect user feedback after every round of adaptation to adjust the weights of constraints. Another approach relies on user feedback to switch between "variants" with associated utility function weights, depending on the current usage context [9]. Our work focuses on eliciting priorities to define utility function weights based on a consensus between multiple stakeholders. As part of future work, we aim to also consider different usage contexts/scenarios in our method.

5   Discussion, Conclusion, and Future Work

In this paper, we presented a method to define utility functions for self-adaptive systems by eliciting and negotiating the priorities of quality attributes. The method is based on the AHP for the pairwise comparison of quality attributes and a consensus-building approach using the Delphi technique. We plan to study the method's application on existing systems in a multiple case study. The method is at an early stage of investigation and needs to be refined further.

For instance, the current method assumes that stakeholders are aware of relevant and measurable quality attributes that can be expressed in functions. For individual quality attributes, techniques are needed to define the intervals of values that are considered "good enough" or "insufficient." Our method will also be extended to explicitly consider hard constraints. Criteria mandated by law (e.g., security or safety constraints) cannot be traded against other preferences. Moreover, the utility of a system strongly depends on its context (e.g., current tasks, security attacks, or faults), which should also be considered, so that human input for utility function definition can be collected when needed and the utility function can be evolved over time. Another relevant concern is to ensure that stakeholders do not over-rate quality attributes to counter for others' conflicting preferences, as it is not always possible to assume non-competitive stakeholders.

We envision our method to be integrated into existing approaches, so that multiple stakeholders' preferences and requirements can be more easily elicited, negotiated, and fulfilled. The presented ideas might also be beneficial for artifacts at other levels of abstraction, e.g., the prioritization of goals or requirements. Moreover, we plan to work on explaining utility functions by describing different priorities' impact on the concrete actions of a system. Our vision is to demystify utility functions by providing human stakeholders with lightweight and understandable methods for the definition and refinement of utility functions.

Acknowledgments. This work is supported in part by the Wallenberg AI, Autonomous Systems and Software Program (WASP) funded by the Knut and Alice Wallenberg Foundation, by award N00014172899 from the Office of Naval Research and by the NSA under Award No. H9823018D000. Any views, opinions, findings and conclusions or recommendations expressed in this material are those of the authors and do not necessarily reflect the views of the Office of Naval Research or the NSA.

References

1. Abdennadher, I., Rodriguez, I.B., Jmaiel, M.: A utility-based approach for self-adaptive systems: application to a smart building. In: AICCSA, pp. 76–82 (2018)
2. Butler, S.A., Fischbeck, P.: Multi-attribute risk assessment. In: SREIS 2002 (2002)
3. Cheng, S.W., Garlan, D., Schmerl, B.: Architecture-based self-adaptation in the presence of multiple objectives. In: SEAMS 2006 (2006)
4. Faniyi, F., Lewis, P.R., et al.: Architecting self-aware software systems. In: WICSA 2014, pp. 91–94 (2014)
5. Ghezzi, C., Molzam Sharifloo, A.: Dealing with non-functional requirements for adaptive systems via dynamic software product-lines. In: de Lemos, R., Giese, H., Müller, H.A., Shaw, M. (eds.) Software Engineering for Self-Adaptive Systems II. LNCS, vol. 7475, pp. 191–213. Springer, Heidelberg (2013). https://doi.org/10.1007/978-3-642-35813-5_8
6. Heaven, W., Sykes, D., Magee, J., Kramer, J.: A case study in goal-driven architectural adaptation. In: Cheng, B.H.C., de Lemos, R., Giese, H., Inverardi, P., Magee, J. (eds.) Software Engineering for Self-Adaptive Systems. LNCS, vol. 5525, pp. 109–127. Springer, Heidelberg (2009). https://doi.org/10.1007/978-3-642-02161-9_6

7. Hsu, C.C., Sandford, B.A.: The Delphi technique: making sense of consensus. Pract. Assess. Res. Eval. **12**(1), 10 (2007)
8. Inverardi, P., Mori, M.: A software lifecycle process to support consistent evolutions. In: de Lemos, R., Giese, H., Müller, H.A., Shaw, M. (eds.) Software Engineering for Self-Adaptive Systems II. LNCS, vol. 7475, pp. 239–264. Springer, Heidelberg (2013). https://doi.org/10.1007/978-3-642-35813-5_10
9. Kakousis, K., Paspallis, N., Papadopoulos, G.A.: Optimizing the utility function-based self-adaptive behavior of context-aware systems using user feedback. In: Meersman, R., Tari, Z. (eds.) OTM 2008. LNCS, vol. 5331, pp. 657–674. Springer, Heidelberg (2008). https://doi.org/10.1007/978-3-540-88871-0_46
10. Kendall, M.G., Smith, B.B.: The problem of m rankings. Ann. Math. Statist. **10**(3), 275–287 (1939)
11. Ossadnik, W., Schinke, S., Kaspar, R.H.: Group aggregation techniques for analytic hierarchy process and analytic network process: a comparative analysis. Group Decis. Negot. **25**(2), 421–457 (2016). https://doi.org/10.1007/s10726-015-9448-4
12. Poladian, V., Sousa, J.P., Garlan, D., Shaw, M.: Dynamic configuration of resource-aware services. In: ICSE 2004 (2004)
13. Runeson, P., Höst, M.: Guidelines for conducting and reporting case study research in software engineering. Empir. Softw. Eng. **14**(2), 131–164 (2009). https://doi.org/10.1007/s10664-008-9102-8
14. Saaty, R.: The analytic hierarchy process–what it is and how it is used. Math. Modell. **9**(3), 161–176 (1987)
15. Salehie, M., Tahvildari, L.: Towards a goal-driven approach to action selection in self-adaptive software. Softw. Pract. Exp. **42**(2), 211–233 (2012)
16. Sawyer, P., Bencomo, N., et al.: Requirements-aware systems: a research agenda for RE for self-adaptive systems. In: RE 2010, pp. 95–103 (2010)
17. Song, H., Barrett, S., Clarke, A., Clarke, S.: Self-adaptation with end-user preferences: using run-time models and constraint solving. In: Moreira, A., Schätz, B., Gray, J., Vallecillo, A., Clarke, P. (eds.) MODELS 2013. LNCS, vol. 8107, pp. 555–571. Springer, Heidelberg (2013). https://doi.org/10.1007/978-3-642-41533-3_34
18. Sousa, J.P., Balan, R.K., Poladian, V., Garlan, D., Satyanarayanan, M.: User guidance of resource-adaptive systems. In: ICSOFT 2008, pp. 36–44 (2008)

Risk-Driven Compliance Assurance for Collaborative AI Systems: A Vision Paper

Matteo Camilli[1]($\boxtimes$), Michael Felderer[2], Andrea Giusti[3],
Dominik Tobias Matt[1,3], Anna Perini[4], Barbara Russo[1], and Angelo Susi[4]

[1] Free University of Bozen-Bolzano, Bolzano, Italy
{mcamilli,dmatt,brusso}@unibz.it
[2] University of Innsbruck, Innsbruck, Austria
michael.felderer@uibk.ac.at
[3] Fraunhofer Italia Research, Bolzano, Italy
andrea.giusti@fraunhofer.it
[4] Fondazione Bruno Kessler (FBK), Trento, Italy
{perini,susi}@fbk.eu

Abstract. *Context and motivation.* Collaborative AI systems aim at working together with humans in a shared space. Building these systems, which comply with quality requirements, domain specific standards and regulations is a challenging research direction. This challenge is even more exacerbated for new generation of systems that leverage on machine learning components rather than deductive (top-down programmed) AI.

Question/problem. How can requirements engineering, together with software and systems engineering, contribute towards the objective of building flexible and compliant collaborative AI with strong assurances?

Principal idea/results. In this paper, we identify three main research directions: automated specification and management of compliance requirements, and their alignment with assurance cases; risk management; and risk-driven assurance methods. Each one tackles challenges that currently hinder engineering processes in this context.

Contributions. This vision paper aims at fostering further discussion on the challenges and research directions towards appropriate methods and tools to engineer collaborative AI systems in compliance with existing standards, norms, and regulations.

Keywords: Compliance requirements · Compliance cases · Collaborative AI systems · Machine Learning · Risk management

1 Introduction

Collaborative AI systems (CAIS) are robotic systems that work with humans in a shared physical space to reach common goals. To achieve flexibility and accommodate changing needs, the upcoming generation of CAIS heavily rely on Machine Learning (ML) components to mimic human perception skills (e.g., visual perception, speech recognition, or conversing in natural language) as

© Springer Nature Switzerland AG 2021
F. Dalpiaz and P. Spoletini (Eds.): REFSQ 2021, LNCS 12685, pp. 123–130, 2021.
https://doi.org/10.1007/978-3-030-73128-1_9

well as learn from humans how to carry out specific tasks by demonstration. Thus, ML-equipped CAIS yield bidirectional human-robot collaboration. For this reason, they must satisfy quality criteria including appropriate behavior with respect to social rules, domain specific standards and laws for certification. Furthermore, such systems often run in dynamic and uncertain environments that make it difficult providing strong assurances of *compliance* [4].

In this paper, we reflect on how research in Requirements Engineering (RE) of software and systems can contribute to define suitable methods for building ML-equipped CAIS (henceforth referred again to as CAIS for the sake of simplicity). We believe that a RE perspective can help reasoning on the trade-off between opportunities and risks deriving from the usage of ML-based solutions in this context. Moreover, existing practices for trust-based human-robot interactions [12] shall be evaluated and eventually revisited through novel RE methods to deal with the assurance of learning agents, in which interactions are driven by ML components. In this setting, we envision risk as a first class concern and propose three research directions to investigate over risk-driven engineering processes that leverage on continuous feedback from empirical evidence collected at run-time, in a closed-loop setting with the surrounding environment, in order to verify semantically meaningful properties through suitable assurance methods.

The RE community recognizes that appropriate assurance methods for compliance requirements (e.g., defined on human-robot collaboration standards, such as the ISO/TS 15066[1]) need to be defined. More generally, research on RE for AI-based system is considered a relevant and timely topic in recent RE conferences[2] and in dedicated workshops, such as RE4AI at REFSQ[3], as well as in European initiatives (e.g., AI4EU platform[4]). Evidence is also provided by recent surveys reporting the urgent need for effective RE processes [14] and verification methods [11] for "intelligent" components as well as AI systems. An analysis of the RE characteristics for systems that include ML components are reported in [14] whereas the work in [2] discusses challenges and desiderata for increasing their level of assurance. This latter work emphasizes that existing assurance methods for AI systems in general and CAIS in particular are not linked to compliance requirements and possible risks.

The rest of the paper is organized as follows. In Sect. 2, we introduce an illustrative example of CAIS. We discuss the key challenges in Sect. 3. Then, we elaborate on our envisioned approach and research directions in Sect. 4. Finally, Sect. 5 concludes the paper.

2   Illustrative Example

We introduce our vision of the problem by means of a case from the Industry 4.0 domain taken from [9] that exemplifies the high risks for human safety as

---

[1] https://www.iso.org/standard/62996.html.
[2] https://requirements-engineering.org/.
[3] https://sites.google.com/view/re4ai/home.
[4] http://ai4eu.eu.

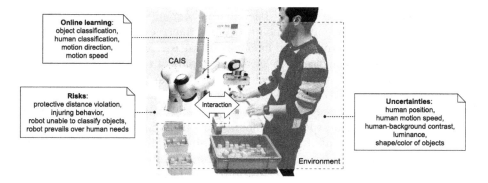

Fig. 1. Illustrative CAIS example in a closed-loop with the surroundings.

well as relevant dependability concerns. Figure 1 illustrates such example, where an automated controller of a robotic arm attempts to detect and classify objects (e.g., by color and shape) on a conveyor belt and actuates the proper movements to pick and move the object into the right bucket. The system includes a controller, an actuated mechanical system (i.e., robotic arm), and a camera sensor along with a visual perception ML component for classification. This ML component learns on structured heterogeneous data sources associated with *features* (e.g., shape and color of an object) and yields category labels as output. The training and the validation is performed iteratively online by a human operator. The operator collaborates with the robot in order to validate its own actions through gestures.

The operator collaborates with the robot in order to supervise the correct transfer of the desired sorting skill to the robot and can intervene through gestures when corrections are required. The safety control approach of this example is implemented by considering the *speed and separation monitoring* of the standard ISO/TS 15066 for collaborative robotics. Here, a protective separation distance between the human and the robot is checked online using *safety zones*. The dimension of such zones is dynamically adapted based on the robot motion. Fast motions of the robot can generate safety zones which may be large and therefore negatively affect the realization of collaborative operations. Assuring a safe and successful collaboration in cases in which the robot motions are learnt from humans yields challenges discussed in the next section.

3   Research Challenges

By collaborating with researchers in CAIS engineering, we started to elicit major challenges that are not fully addressed by existing approaches. In this section, we discuss them in light of the key characteristics of CAIS, by focusing on the objective of providing comprehensive, ideally provable, evidence that CAIS exhibit dependable behavior within their viability (due to continuous learning).

CH1 *Uncertain Environment:* The environment in which CAIS operate is often complex with substantial amount of uncertainty even in scenarios where interacting agents (both robots and humans) are known. For instance, the human operator in the running example (Sect. 1) is key part of the environment. Thus, assurance methods for CAIS shall deal with inherent variability and uncertainty in human behavior. As an example, the human operator may unpredictably move in a forbidden area and the robotic arm must be able to react in a safe way by enforcing the speed and separation monitoring. In addition to humans, other environment variables can influence the perception capability of the ML components, as shown in Fig. 1. For instance, low luminance might lead to decrease the ability of classifying the human operator in specific locations of the shared space. In this case, assurance could leverage on probabilistic approaches by assuming specific distributions of the environment factors. Nonetheless, underlying distributions are often only estimates and do not represent precisely the environment behavior [5].

CH2 *Adequacy of Standards:* Existing standards in the domain of CAIS pose challenges in realizing flexible automation [8]. Difficulties arise from frequently changing production environments and potentially unknown *a-priori* robot motions. These issues become particularly severe when robotic systems swiftly adapt to different task demands by learning from humans using ML components [3]. In fact, existing standards do not specifically refer to CAIS able to learn from humans (e.g., through ML components). For instance, in programming by demonstration [8], fast human motions can induce in turn fast robot motions. Thus, to enforce speed and separation monitoring, the dimension of the safety zones might become large and therefore negatively affect the realization of a successful collaboration. Operational phases can suggest feedback to existing norms and standards that are currently not aligned with practical needs of humans in the context of CAIS. Without clear regulation on what the ML component shall (or shall not) learn, we could put in production CAIS that eventually break trust and prevail over human needs.

CH3 *Partial and Evolving Specifications:* Strong compliance assurances usually rely on precise, rigorous, or even formal description of the behavior of the target system. This is anything but trivial in the human-robot collaboration domain and even exacerbated when the robot makes use of ML components. In fact, the data is often the only available "ground truth" of correct behavior for ML components. Available data can only partially represent the correct behavior of CAIS. In our running example (Sect. 2), the model of the operator's behavior built on available data might not cover an unforeseen movement toward forbidden unsafe areas of the human agent. Furthermore, such behavior constantly changes due to the learning skills of the system. Since CAIS learn from new execution scenarios, design-time specifications must either account for future changes or be incrementally refined online, as the system evolves.

CH4 *Top-Down/Bottom-Up Duality:* There exists a fading boundary between the two approaches in the emerging assurance methods tailored to CAIS.

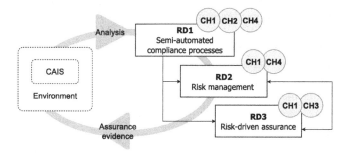

Fig. 2. Research directions RD (Sect. 4) and related challenges CH (Sect. 3).

For instance, refinement of top-down decomposition of requirements from standards interleaved with bottom-up analysis of human needs. Another example is top-down partial specification interleaved with bottom-up run-time assurance evidence. Therefore, "traditional" top-down specifications of requirements shall coexist with partial/incomplete or example-based specifications (i.e., examples of good/bad behaviors). In our running example, a top-down safety compliance requirement could be "no injuring behaviors with probability 99.99%" whereas the ML visual perception component is built bottom-up on the positive and negative examples (i.e., injured humans), which may not be available. Overall, the top-down/bottom-up duality nature calls for revisited methods built with awareness and endowed with the ability to interleave the two facets.

4   Research Roadmap

In this section, we introduce a research roadmap to address the challenges discussed above. Figure 2 shows the high level schema of this roadmap that outlines the Research Directions (RD), and the challenges (CH) that they face. We are following it in our collaboration with Fraunhofer Italia ARENA[5] where we will have the opportunity to validate results in real world CAIS in the industrial manufacturing domain.

RD1 *Semi-automated Compliance Processes:* The focus of this RD is on supporting semi-automated derivation of compliance requirements of CAIS from norms and their embedding into proper assurance cases. Requirements elicitation should rest on the continuous analysis of the uncertain environment (CH1) and, as a consequence, the management of possible risks. Multi-paradigm approaches (e.g., goal-oriented techniques [10]) could be exploited to capture different facets of relevant standards and norms regulating CAIS. A promising approach here is the extension of existing modelling techniques by enabling (semi)-automated creation of norm models from domain standards and rules described using natural language (e.g., by exploiting NLP [13]). Furthermore, domain expert knowledge

---

[5] https://www.fraunhofer.it/en/focus.html.

must be considered to support norms evolution based on the feedback and data from the field operation (CH2). Quality criteria for CAIS and a set of compliance requirements (e.g., *coverage* metrics [13]) represents another immature research direction. Namely, the definition of techniques to automate quality assurance processes as well as providing recommendations to engineers for improving the compliance requirements themselves (CH4) require further investigation.

RD2 *Risk Management:* A successful strategy to cope with uncertainty (CH1), while dealing with safety in particular and dependability concerns in general, is the adoption of a risk management perspective. Uncertainty in CAIS leads to risks that must be identified, analyzed, monitored, and mitigated. Risk models could be adopted to quantify the risk of breaking compliance requirements. The notion of risk is a combination of the likelihood and impact of negative events. In our context, we see the likelihood as the probability that during execution, specific characteristics of the environment (i.e., domain features) cause problems to the ML component (e.g., misclassifications) that might break compliance requirements [7]. Data and domain models will be used to describe constraints, partitions and ranges of domain features (e.g., possible positions of the robotic arm in our running example) and constraints on the underlying data source (e.g., reliability of sensors), respectively. Risk analysis has the potential of quantifying existing risks by integrating probability and impact functions in order to prioritize risks and related compliance requirements that need special attention during the compliance assurance processes. This should inform the top-down and bottom-up compliance requirements management strategies in the learning and operative collaborative activities (CH4). The results of the assurance methods can be used to further bring the overall residual risk below acceptable levels, as defined by proper thresholds associated with assurance cases. Here, there exists the urgent need of novel adequacy criteria prescribing meaningful thresholds for residual risk in the context of CAIS.

RD3 *Risk-Driven Assurance:* The risk models developed in RD2 have the potential to drive the automated creation of semantic labels associated with uncertain operating conditions that yield risks (CH1). Prioritized assurance cases and testing scenarios can then be derived from such operating conditions. The comprehensiveness of testing activities, should be then assessed by means of appropriate coverage metrics as discussed in the context of RD1. Further assurance can be provided by using incremental refinement of partial/incomplete specifications (CH3) through verification activities. Partial knowledge captured by risk models can be used to sample execution scenarios associated with high risk. Then run-time data can provide evidence about system compliance that can be used in turn to update the prior knowledge as defined by the risk model. As an example, *falsification* techniques [1,6] (traditionally applied in the context of cyber-physical systems) enhanced with awareness on risks have the potential of driving a CAIS towards compliance issues. In our running example, we could, for instance, falsify a safety compliance property requiring a minimum distance between the robot and the human operator.

5   Conclusion

In this paper, we provided a reflection on how research in the discipline of RE, and software/systems engineering can contribute towards the objective of building effective ML-equipped CAIS with strong, ideally provable compliance assurances. Major challenges that hinder our ultimate goal are discussed to rise awareness and call for contributions from the research community. To deal with these challenges, we designed a research road-map towards the definition of compliance processes and requirements, risk management for ML-equipped CAIS, and risk-driven assurance.

References

1. Abbas, H., Fainekos, G., Sankaranarayanan, S., Ivančić, F., Gupta, A.: Probabilistic temporal logic falsification of cyber-physical systems. ACM Trans. Embed. Comput. Syst. **12**(2s), 1–30 (2013)
2. Ashmore, R., Calinescu, R., Paterson, C.: Assuring the machine learning lifecycle: desiderata, methods, and challenges (2019). https://arxiv.org/abs/1905.04223. Accessed Nov 2020
3. Billard, A.G., Calinon, S., Dillmann, R.: Learning from humans. In: Siciliano, B., Khatib, O. (eds.) Springer Handbook of Robotics, pp. 1995–2014. Springer, Cham (2016). https://doi.org/10.1007/978-3-319-32552-1_74
4. Breaux, T.D., Vail, M.W., Anton, A.I.: Towards regulatory compliance: extracting rights and obligations to align requirements with regulations. In: 14th IEEE International Requirements Engineering Conference (RE), pp. 49–58. IEEE (2006)
5. Camilli, M., Russo, B.: Model-based testing under parametric variability of uncertain beliefs. In: de Boer, F., Cerone, A. (eds.) SEFM 2020. LNCS, vol. 12310, pp. 175–192. Springer, Cham (2020). https://doi.org/10.1007/978-3-030-58768-0_10
6. Dreossi, T., et al.: VERIFAI: a toolkit for the formal design and analysis of artificial intelligence-based systems. In: Dillig, I., Tasiran, S. (eds.) CAV 2019. LNCS, vol. 11561, pp. 432–442. Springer, Cham (2019). https://doi.org/10.1007/978-3-030-25540-4_25
7. Foidl, H., Felderer, M.: Risk-based data validation in machine learning-based software systems. In: Proceedings of the 3rd ACM SIGSOFT International Workshop on Machine Learning Techniques for Software Quality Evaluation, pp. 13–18 (2019)
8. Giusti, A., et al.: Flexible automation driven by demonstration: leveraging strategies that simplify robotics. IEEE Robot. Autom. Mag. **25**(2), 18–27 (2018)
9. Giusti, A., et al.: Kollaborative robotik - maschinelles lernen durch imitation. Industrie 4.0 Management, pp. 43–46 (2019)
10. Ishikawa, F., Matsuno, Y.: Evidence-driven requirements engineering for uncertainty of machine learning-based systems. In: 2020 IEEE 28th International Requirements Engineering Conference (RE), pp. 346–351. IEEE (2020)
11. Ishikawa, F., Yoshioka, N.: How do engineers perceive difficulties in engineering of machine-learning systems? - questionnaire survey. In: 2019 IEEE/ACM Joint 7th International Workshop on Conducting Empirical Studies in Industry (CESI) and 6th International Workshop on Software Engineering Research and Industrial Practice (SER&IP), pp. 2–9. IEEE (2019)

12. Rahman, S.M., Wang, Y., Walker, I.D., Mears, L., Pak, R., Remy, S.: Trust-based compliant robot-human handovers of payloads in collaborative assembly in flexible manufacturing. In: 2016 IEEE International Conference on Automation Science and Engineering (CASE), pp. 355–360. IEEE (2016)
13. Torrea, D., et al.: An AI-assisted approach for checking the completeness of privacy policies against GDPR. In: 28th IEEE International Requirements Engineering Conference, RE 2020, Zurich, Swiss, 31 August–4 September 2020, pp. 136–146 (2020)
14. Vogelsang, A., Borg, M.: Requirements engineering for machine learning: perspectives from data scientists. In: 2019 IEEE 27th International Requirements Engineering Conference Workshops (REW), pp. 245–251. IEEE (2019)

From Software to Systems and Services

Requirements Engineering in the Planning Phase of a Software Ecosystem

Kati Saarni[✉] and Marjo Kauppinen

Department of Computer Science, Aalto University, Espoo, Finland
marjo.kauppinen@aalto.fi

Abstract. [**Motivation**] Companies are building software ecosystems to gain competitive advantage by developing digital services together for customers. The planning phase of the software ecosystem can, however, be challenging. [**Question**] The goal of this study was to analyze what the role of requirements engineering (RE) was in the planning phase of a small-sized software ecosystem. The case study was conducted by interviewing representatives of all six actors of the ecosystem and analyzing material from the 12 planning workshops. [**Results**] The paper describes the conceptualization process of digital services the actors used during the planning phase. This process contained a flow of tasks from a vision and objectives of the software ecosystem to a go/no-go decision on the development of a Minimum Viable Product (MVP). One key characteristic of the conceptualization process was to have traceability from the prioritized functionalities of the MVP to a value proposition, target customer groups and customer paths of digital services and further to the vision and objectives of the software ecosystem. [**Contribution**] The paper provides knowledge on how actors can start building a software ecosystem together from a business perspective. In addition, it addresses the importance of RE to link the business view to the development of the MVP of digital services in the software ecosystem.

Keywords: Software ecosystem · Requirements engineering · Business view · Conceptualization process · Planning

1 Introduction

Companies are building business ecosystems together to reach competitive advantage in the markets. The business ecosystem concept was proposed by Moore [18] in the 1990s. The business ecosystem where digital services are developed and provided can be considered as a software ecosystem (SECO) [11]. It can be defined thus: *"a set of actors interact with a shared market, develop software and services together and operate through the exchange of information, resources and artifacts"* [11]. The creation of the ecosystem starts from a planning phase, where a basic paradigm of the ecosystem and how value will be created and shared need to be determined [18]. The software ecosystem enables the actors to build a broader set of services than one actor can do by its own [10]. In addition, it allows the actors to better address customer needs, as it can bring a

© Springer Nature Switzerland AG 2021
F. Dalpiaz and P. Spoletini (Eds.): REFSQ 2021, LNCS 12685, pp. 133–148, 2021.
https://doi.org/10.1007/978-3-030-73128-1_10

diverse set of capabilities and innovation to the solution quickly [5]. It is also possible to tolerate risk through cost-sharing [10].

Manikas and Hansen [15] point out the importance of studying existing and real software ecosystems. Understanding different perspectives of specific types of software ecosystems can provide results which can then be applied to other software ecosystems [16]. This will enable repeatability and theory confirmation [16].

Earlier studies have examined requirements engineering (RE) in software ecosystems [e.g. 16, 24, 25]. The focus of these studies has varied across RE activities [24] or ecosystem lifecycles in small- to large-sized software ecosystems [16, 24]. RE-related research concentrating on the planning phase of small- and medium-sized software ecosystems has had only little attention in recent studies [16, 24].

In our previous paper, we identified the main activities and challenges in the planning phase of a software ecosystem [21]. However, we are interested in gaining a deep understanding of RE when actors are building digital services together in the planning phase of the software ecosystem. Therefore, we extended the quality analysis of the collected data to answer *the following research question: What is the role of RE in the planning phase of a software ecosystem?* The main contribution of this study is that it provides information for practitioners on how they can start building a software ecosystem. In addition, the paper addresses the importance of RE in linking the business view to the development of the MVP of the digital services in the software ecosystem.

The rest of the paper is organized as follows. Section 2 summarizes the main concepts of software ecosystems and gives an overview of existing RE research on software ecosystems. The qualitative research method of the study is described in Sect. 3. The conceptualization process of digital services is described in detail in Sect. 4. The results are discussed and the answer to the research question presented in Sect. 5. Finally, the paper concludes and points to future research.

2   Related Work

2.1   Overview of Software Ecosystems

In the 1990s, Moore [18] proposed the concept of the business ecosystem, concentrating on how the economic community worked and the interactions between companies, their business environments and business opportunities. A software ecosystem is a subset of a business ecosystem and the literature contains many definitions of the SECO [e.g. 1, 9, 11, 12]. The main common characteristic of all these definitions of the SECO is the use of software, which differentiates SECOs from other ecosystem types. In this paper, we use the definition by Jansen et al. [11] of a SECO: *"a set of actors functioning as a unit and interacting with a shared market for software and services, together with the relationships among them. These relationships are frequently underpinned by a common technological platform or market and operate through the exchange of information, resources and artifacts"*.

Software ecosystems can be classified through four factors: base technology, coordinators, market extensions and accessibility [12]. This means that a software ecosystem is always underpinned by a software platform, a software service platform, or a software standard, and it can be a privately owned or owned by a community [12]. In addition, there are several scenarios for how the software ecosystems can be available in the markets e.g. commercial market extensions mean the owner(s) will take all profits [12]. The accessibility possibilities can be open source, screened but free, and paid [12].

Participants in software ecosystems can be called actors and can have different roles e.g. keystone actors, dominators, hub landlords and niche players [10]. The software ecosystem is usually governed [12] and the digital service development is often led by one or more keystone actors [10]. An actor may have one or more roles in the software ecosystem [13], and their role may also change during the ecosystem's life cycle [17]. The lifecycle of an ecosystem consists of phases [10, 18], where an early phase of the ecosystem is for example referred to as its birth [18] or emerging [10] phase. We call the first phase of building software ecosystems the planning phase [21].

Software ecosystems can be characterized in several ways, for example information about the owning companies, number of participants, main customers as well as how much business has been created through the software ecosystem [11]. Campbell and Ahmed [4] proposed a three-dimensional view of the development of software ecosystems, consisting of the business, architecture and social views. The business view includes activities where customer expectations and competitive advantage are reached by a business vision, innovation and strategic planning [4]. The architecture view focuses, for example, on software architecture solutions for software ecosystems [1] whereas the social view looks at, for example, how the actors negotiate during the planning of the digital services development to achieve the goals [8].

In addition, we have identified in our previous study five activities in the planning phase of the software ecosystem: 1) definition of a vision and objectives, 2) selection of actors, 3) definition of a governance model, 4) conceptualization of digital services, and 5) definition of a business model [21]. Other studies also pointed out that definition of a vision and objectives [10, 18, 20] and definition of the roles of actors [7, 10] are important activities in the planning phase.

2.2 Requirements Engineering in Software Ecosystems

Earlier studies have considered requirements engineering (RE) from different perspectives in software ecosystems [e.g. 8, 16, 23–25]. Vegendla et al. [24] map previous studies to RE activities. The previous studies have also been directed toward different phases during the lifecycle of a software ecosystem [16, 24]. In addition, the size of the software ecosystems varied from small- and medium-sized [e.g. 23] to large-scale ecosystems [e.g. 8, 27].

Goal modeling [27] and a definition of a common value proposition [19] as well as requirement negotiation [8] and prioritization [23] are activities which occur in the planning phase of a software ecosystem. Yu and Deng [27] proposed a modeling approach for achieving the strategic goals of each actor in a large-scale software ecosystem. Pichlis et al. [19] studied small software ecosystems and emphasized a need for a common value proposition. Fricker [8] proposed a model based on negotiation and network theory for analyzing and designing the flow of requirements through a large-scale software ecosystem. Valenca et al. [23] studied small- to medium-sized companies during the planning of a software ecosystem. They identified a need for strong strategic alignment and difficulties in prioritizing the most valuable features.

Villeda et al. [25] identified an iterative process to achieve a first version of the digital services in the software ecosystem. The first step in this process is the definition of a preliminary software ecosystem concept. End-user roles, business strategy and needed software services from actors to accomplish the strategy are included to the concept [25].

Bosch and Sijtsema [1] reported a lack of connection between business and engineering process in a large-scale software ecosystem. Schultis et al. [22] reported challenges where the actors have different requirements based on their business objectives, and if all the actors are involved in the architectural decision-making, it takes time to reach a common agreement on the architecture. These results have occurred more in the development phase, but relationships to the planning phase exist.

3   Research Methods

3.1   Research Question

The goal of this study was to understand the role of the requirements engineering when the actors were building digital services together in the planning phase of a software ecosystem. The research question of the study is defined as follows: **What is the role of requirements engineering in the planning phase of a software ecosystem?**

3.2   Description of Case SECO

The role of requirements engineering in the planning phase of one Finnish software ecosystem (called Case SECO in this paper) was investigated in this study. The aim of Case SECO was to provide digital services for new entrepreneurs. Before the actual planning phase of Case SECO, three cooperating actors had recognized that there was a need in the market for comprehensive digital services. Therefore, they were interested in co-creating a targeted offering for them. They recognized that creating this kind of digital service offering requires a set of companies developing it together. A software ecosystem was recognized as a suitable model for this kind of cooperation. The actors started to gather appropriate companies. Based on the preliminary discussions with them, potential companies were selected.

The planning of the software ecosystem took place from February to June 2018 and was performed through 12 workshops. In the beginning, there were five actors, and the sixth actor joined the planning phase in the eighth workshop. The actors represented five different business sectors: two actors were categorized as small- and medium-sized companies and four were large companies. One to three people from each actor attended the workshops and all members actively participated in defining the vision and objectives of the software ecosystem and a set of digital services to be developed during the workshops. One actor took the role of facilitating the planning phase because it had previous experience of ecosystem creation and knowledge of digital services development. The planning was done in an iterative manner.

During the planning phase, the actors agreed that all of them had a keystone player's role and were in an equal position with each other in decision-making. An advisory board was set up consisting of one member of each of the actors of this planning phase. The advisory board in the was the highest decision-making governance body, to enable the planning of the ecosystem and steer the planning of the digital services. The roles and responsibilities, limitations, cost-sharing principles, rules for co-operation and business model were described in the rule book, which was the main guiding document for the governance of Case SECO.

Case SECO can be classified to be a closed and privately-owned software ecosystem. All six actors owned equal parts of the software ecosystem. In addition, they decided to share the costs of the planning and development of the software ecosystem equally. They also agreed that each actor would get the profit that came from their own offering through the digital services. The actors decided to keep Case SECO closed during the planning phase and not to take on new members. However, they agreed that new members would be welcome later if their offering is suitable for the end-users of the digital services. The actors included rules for joining and possible roles for new actors in the rule book. The digital services fulfilled the vision and objectives of Case SECO, which the actors defined together. The digital services were executed through one software platform, which was developed by a one external development team. Each actor's offering showed up as solid digital services for the end-users.

The development phase of the digital services started in July 2018 and the first version was launched in July 2019. Currently the digital services are in the continuous development phase.

3.3   Research Process

This qualitative research was performed using a case study research method [26]. A descriptive approach for the case study was used to describe a single case in depth. We applied the coding and code comparison guidelines of grounded theory to analyze the data [6]. The grounded theory method was selected for the analysis because it offers systematic and flexible guidelines for analyzing qualitative data [6]. The research process was first presented in our publication [21], where open coding was used to analyze the data. In this paper, we further analyzed the data by applying axial coding [6]. Figure 1 shows a timeline of the phases of Case SECO and the main research activities of this study.

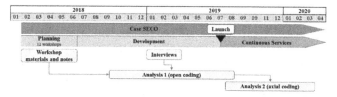

Fig. 1. Timeline of the phases of Case SECO and the main research activities.

The answers to the research question of this study were based on the workshop materials from the planning phase and the results of semi-structured interviews performed in January and March 2019. The workshops had a predefined agenda, but other topics were also covered. The length of the workshops varied from 1 to 4.5 h. The workshop materials included presentations, notes and all additional material which was delivered to the actors during the planning. The actors prepared the workshops together with pre-agreed responsibilities. The facilitator took care of the notes, which included decisions and work items with responsibilities in each workshop.

The interviews were designed following the guidelines from Boyce and Neale [2]. The themes of the interviews covered main topics related to software ecosystem creation. All the six actors, that had a keystone role in Case SECO and were participants on the advisory board in the planning phase, were interviewed. All the interviewees had over 15 years of work experience and had extensive knowledge of their company's business and its development. Only one had previous experience of planning ecosystems together with other actors. Table 1 summarizes the interviewed actors.

Table 1. Summary of the interviewees.

Business sector	Company size	Role in the company	Ecosystem experience
Insurance	Large	Business development director	No
Pension insurance	Large	Business development director	No
Telecommunication	Large	Business director	No
Financial and accounting	Medium	Chief executive officer	No
Financial and accounting	Medium	Business development director	No
Information and communication	Large	Principal consultant	Yes

The interviews were conducted in Finnish, because Finnish was the mother tongue of all the interviewees, and we wanted to collect as rich data as possible. The length of each interview varied from 25 min to 55 min. The interviews were recorded and transcribed by a professional external organization.

Open coding [6] was used to analyze the data and define the main activities and challenges in the planning phase of a software ecosystem. The results of this analysis are reported in our already published paper [21]. In this paper, we further analyzed the tasks and challenges of the conceptualization activity of digital services by applying axial coding from the grounded theory method [6]. Axial coding was used to identify the relationships between the tasks of the conceptualization activity. The outcome of this analysis is the visualization and description of the conceptualization process that was used in Case SECO. This process was further analyzed from the RE perspective. First, we identified important RE activities the actors applied during the conceptualization process of the digital services. After this, we analyzed the benefits and relationships of the identified RE activities. Finally, we categorized the important RE activities of the conceptualization process according to requirements elicitation, analysis, representation and validation. This led us to understand the role of RE in the planning phase of a software ecosystem.

4 Results

4.1 Overview of the Conceptualization Process of Digital Services

Figure 2 summarizes **the conceptualization process of digital services in the planning phase** of Case SECO. The six actors of the software ecosystem defined a shared vision and main objectives of Case SECO in the first planning workshop before the conceptualization process started. The shared vision and main objectives provided important information for the conceptualization process and they were also adjusted during the conceptualization. The process consisted of two sub-processes: **a high-level conceptualization** and **a detailed conceptualization.**

The first sub-process, the high-level conceptualization of digital services, included three tasks: benchmarking existing similar digital services, definition of a value proposition and definition of target customer groups and customer paths. The actors of the software ecosystem did these three tasks in parallel and in an iterative manner in the first three workshops.

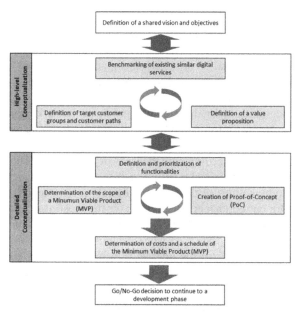

Fig. 2. The conceptualization process of digital services in the planning phase.

The second sub-process, the detailed conceptualization of digital services, consisted of four tasks: definition and prioritization of functionalities, creation of a Proof-of-Concept, determination of a Minimum Viable Product (MVP) and determination of the costs and schedule of the MVP. The actors executed the first three tasks in an iterative manner and in parallel in the third to seventh workshops. In addition, minor iterations were also done in the ninth and eleventh workshops. These three tasks provided enough information to define the costs and schedule for the MVP.

After the conceptualization process, the actors saw that they had enough information to make a go/no-go decision on continuing to the development phase. Although the actors understood that the estimated costs and the schedule might change after the requirements had been defined in more detail, the actors decided to continue to the development phase.

4.2 High-Level Conceptualization of Digital Services

The high-level conceptualization contained three very closely related tasks: benchmarking existing similar digital services, definition of a value proposition and definition of target customer groups and customer paths.

The **benchmarking of existing digital services** included a market review and reviewing five existing similar digital services. The market review included statistics about companies already established in Finland, for example, quantity, sizes, forms, industry and business areas. In addition, the gender, age and geographical location of new entrepreneurs were analyzed. This information was gathered from different statistics services. The market review provided information about the market potential, and it helped to understand potential target customer groups and customer paths.

The three main findings from the review of existing similar digital services were: 1) there was not much automation on processes, 2) a lot of information about establishing a company and entrepreneurship was available, but the language was quite bureaucratic and difficult to understand, and 3) the existing digital services were quite expensive. In addition, the actors realized that each existing digital service provided services at some specific point of the entrepreneur's lifecycle and solutions that support the whole lifecycle from the beginning seemed to be missing. The benchmarking of existing digital services provided important information for the definition of a value proposition, target customer groups and customer paths.

When the actors were proceeding with the tasks of the high-level conceptualization, they were also expanding the benchmarking to gain more information. For example, they studied the functionalities of the existing digital services to see if they provided learning materials for entrepreneurs or templates for the most-used business contracts.

Twenty potential end-users were interviewed to understand customer behavior and discover the main pain points they face when establishing a company and starting to be an entrepreneur. The results of the interviews were visualized. An example of the visualization is given in Fig. 3. The results of the interviews impacted on the definition of the value proposition and the definition of the customer paths.

The analysis of existing digital services and, especially, the main pain points of potential end-users provided valuable information for the definition of a **value proposition**. The value proposition of Case SECO was defined to be *"providing believable digital services for end-users, removing uncertainty and enabling carefreeness of end-users through the entrepreneur's lifecycle"*. The value proposition was meant for both defined target customer groups. In addition, the actors saw that it is important that the defined value proposition does not conflict with their company's own values so that they

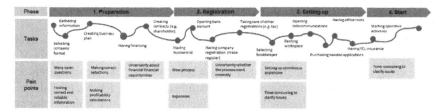

Fig. 3. Visualization of customer behavior and the main pain points of end-users.

can all stand behind it. The value proposition also had an important effect on how the digital services were communicated and marketed to stakeholders.

The benchmarking of existing similar digital services and end-user interviews addressed the need to provide digital services for people who are aiming to be entrepreneurs and those who have recently set up a company and are already entrepreneurs. These two groups were selected to be the **target customer groups** of the digital services. The two target customer groups already represented important customers for each participating actor, and they had good experience of the behavior and needs of these customers.

The main idea when defining the **customer paths** was to support the recognized behavior and remove the main pain points the end-users are currently facing. These were gathered in the potential end-user interviews. The visualization of end-user behaviors was analyzed and further refined to form the customer paths. All of the actors needed to recognize their interests in the defined customer paths even though the actors' specific offering was pointing to only one specific customer path.

Figure 4 shows the relationship between the defined target customer groups and customer paths. The first two customer paths were addressed to persons who are planning to be entrepreneurs: the digital services provide information about available company formats and recommend the most appropriate for the new entrepreneur. They can also set up a company by selecting the company format and receive the needed information during the setting-up process. In addition, the end-users can clearly see where to commit and what the upcoming costs will be. The third customer path serves both target customer groups. The entrepreneur can order tools, services and insurances for operating the company. The ordering process is smooth, and the status and costs are visible to the end-user.

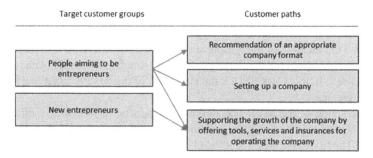

Fig. 4. Relationships between target customer groups and customer paths.

4.3 Detailed Conceptualization of Digital Services

The starting point for the detailed conceptualization of digital services was the defined value proposition, target customer groups and customer paths, which the actors of the software ecosystem did together during the first sub-process. The second sub-process consisted of defining and prioritizing functionalities, creating the Proof-of-Concept (PoC), determining the Minimum Viable Product (MVP) and determining the costs and schedule of the MVP. The actors also had a possibility to shape the results of the high-level conceptualization during the detailed conceptualization process.

The main customer paths were further refined by defining main **functionalities** of the digital services, which can be seen in Fig. 5. In addition, the first **prioritization** for the main functionalities was also done. The defined value proposition guided the definition and prioritization of the main functionalities.

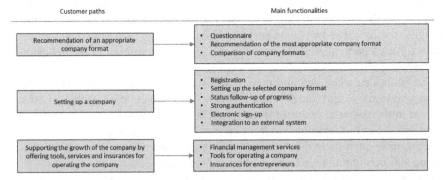

Fig. 5. Relationships between the customer paths and the main functionalities.

The **Proof-of-Concept** was a user interface prototype where the top prioritized functionalities for the main customer paths were drafted and the layout and main interactions defined. The PoC enabled the concrete look and feel of the planned digital services. Potential end-users and the actors tested the PoC and feedback was gathered and discussed in the workshops. The PoC helped the actors to consider the prioritized functionalities, define them in more detail and determine the MVP. During the second sub-process, the PoC was further modified to correspond to the desired set of functionalities and the MVP.

The **Minimum Viable Product** was defined based on the information the PoC provided. The actors agreed that a winning end-user experience should be already in place in the MVP. This meant that everything the end-user sees and how the functionalities work for them should be in place already at the first launch and should not change during the next versions. In this Case SECO, the definition of the MVP on that level was done without disagreements.

After the scope of the MVP was defined, the functionalities were further divided into more detailed requirements. In addition, the main quality requirements were also determined. The first version of a high-level product backlog of the MVP was composed, where all the detailed functional and quality requirements were listed. The actors went through the backlog and considered it against the defined value proposition, target customer groups and customer paths and made some further changes to the prioritization. The result was a prioritized high-level product backlog.

After the MVP had been defined, an overall view of **costs** and a development **schedule** of the MVP were preliminarily determined. The costs included costs for the work needed during the development as well as costs for the technologies and services used. In addition, preliminary costs for continuous services were also estimated. The schedule was based on the estimated work effort for the development and a full time equivalent (FTE) development team. The overall estimation of the costs and the schedule at this point worked as information for decision-making on proceeding. The actors understood that it was an estimation and might change after the definitions of the requirements had become more accurate and the actual development work had started.

4.4   RE Activities of the Conceptualization Process of Digital Services

Table 2 summarizes the set of RE activities that we recommend based on the analysis of the conceptualization process of Case SECO. These RE activities had an important role in the iterative conceptualization process. We categorize the first four (1–4) RE activities to be important requirements elicitation and analysis practices that provide valuable information about the market potential, existing digital services, and especially needs of potential end-users.

The following four RE activities (5–8) can be categorized to be modelling and documentation practices. Before the definition of functionalities of digital services, it was important to define clearly the most relevant target customer groups for whom the digital services will be developed and their customer paths the digital services will support. The starting point for all these modelling practices was to define a value proposition that sets the most important objectives of digital services from the perspective of end-users.

The traceability activity (Activity 9) can be classified to be a very important requirements management practice. Its purpose is to ensure the traceability from the defined functionalities to the value proposition, target customer groups and customer paths of digital services and further to the vision and objectives of the software ecosystem. This traceability activity helped the actors of Case SECO to prioritize functionalities as a team (Activity 10) and define the scope of the MVP (Activity 12).

Prototyping (Activity 11) was one important activity that was used to validate the prioritized functionalities and the definition of the MVP. Organizing workshops throughout the conceptualization process and conducting the RE activities iteratively also supported the validation of the important outcomes of the conceptualization.

4.5   Challenges in the Conceptualization Process

The actors did not face any remarkable challenges during the high-level conceptualization. The main reasons were that the tasks in the high-level conceptualization were executed based on the shared vision and objectives, which the actors had defined together, and the participating actors were very familiar with the substance of the tasks. In addition, the actors did not need to think about costs and schedules yet during the high-level conceptualization.

Table 2. Important RE activities of the conceptualization process of Case SECO.

ID	RE activity	Benefits
1	Conducting a market review	Getting information about the market potential
2	Benchmarking existing services	Understanding the functionalities and weaknesses of the existing digital services
3	Interviewing potential end-users	Gaining a deep understanding of user needs such as their current processes, behavior, and pain points
4	Visualizing customer behavior	Clarifying the current tasks and pain points of potential end-users
5	Defining a value proposition	Deciding the most important objectives of digital services from the perspective of end-users
6	Defining target customer groups	Selecting the most relevant customer groups for whom the digital services will be developed
7	Defining customer paths	Determining targeted customer paths for the defined target customer groups
8	Defining main functionalities	Describing the digital services in more detail and enabling a definition of costs and a schedule for the development of the digital services
9	Ensuring traceability from the vision to functionalities	Helping actors to make compromises and prioritize functionalities and commit to the digital services
10	Prioritizing functionalities as a team	Ensuring that digital services fulfill the defined vision and objectives of the software ecosystem
11	Building a user-interface prototype	Validating the prioritized functionalities and the definition of the MVP
12	Defining the Minimum Viable Product (MVP)	Enabling the launch of a first useful version of the digital service as soon as possible
13	Organizing workshops with all actors	Building trust between all actors and commitment to the created software ecosystem
14	Conducting RE practices iteratively	Enabling enhancements to the outcomes of the conceptualization process

During the detailed conceptualization, the actors faced three main challenges: 1) it was difficult to define the MVP and prioritize functionalities, 2) they had difficulties understanding the needed definition level of the digital services, and 3) they did not have enough substantive knowledge to define common functionalities of the digital services.

The definition of the MVP and prioritization of functionalities was challenging because it required some compromises from the actors. The actors accepted this and understood that the prioritization and thereby the MVP was based on the tasks that they defined very carefully together in the high-level conceptualization and which were strongly based on the shared vision and objectives.

The actors were not sure how detailed the definitions of the requirements should be to gain enough information about the costs and schedule in order to ensure that their go/no-go decision for the proceeding was correct. In addition, the actors considered that they did not have enough substantive knowledge to define the common functionalities (e.g. registering, interactions, security and layout) of the digital services. The detailed conceptualization was an important sub-process and ensured the needed information for the decision-making. Thus, the second sub-process should have included participants who were experts in detailed conceptualization.

5 Discussion

5.1 Requirement Engineering in the Planning Phase of a Software Ecosystem

In this study, we identified the conceptualization process of digital services that was used by the actors in the planning phase of the software ecosystem. One important characteristic of the conceptualization process was that the actors defined the value proposition, the target customer groups and the customer paths of digital services together. It was also important that the value proposition, target customer groups and customer paths were defined based on the vision and objectives of the software ecosystem. This high-level conceptualization ensured that the business view was considered systematically during the planning of the digital services.

Previous research has also highlighted the importance of a business perspective in software ecosystem creation. For example, Yu and Deng [27] stressed the importance of strategic goal definition and Valenca et al. [23] also reported the need for strategic alliances in small- and medium-sized software ecosystems. In addition, Villeda et al. [25] recognized the need for the software ecosystem concept, which included a definition of the business strategy and needed software services from actors to accomplish the strategy.

Our findings support the results of Pichlis et al. [19]. They have reported the need of the common value proposition when developing a software ecosystem. Defining value propositions, target customer groups and customer processes of digital services can also be considered an important part of RE especially when connecting RE to business planning [14].

The analysis of the current pain points of customer processes was one of the critical tasks of the conceptualization process. The purpose of the software ecosystem was to remove these pain points. Gaining a deep understanding of customers' current processes and their problems is also an essential part of requirements elicitation.

During the detailed conceptualization, the actors felt it was challenging to define the MVP and prioritize functionalities. Valenca et al. [23] have also pointed out the challenges in prioritizing product features in a software ecosystem. Requirements prioritization is a challenging RE activity that has been investigated for decades, and researchers have proposed a large number of prioritization methods [3].

The actors needed to make compromises when they prioritized functionalities and defined the MVP during the planning phase of the software ecosystem. Even though the actors found requirements prioritization challenging, they accepted the compromises, because it was very transparent how the MVP was derived from the together defined value proposition, target customer groups and customer paths. Therefore, the actors knew that the MVP fulfilled the vision and objectives of their software ecosystem. This is an example how the actors of the software ecosystem were able to prioritize the requirements of the MVP together without using requirements prioritization methods.

The social view and collaboration were emphasized in the planning phase of this software ecosystem. It was essential that the actors worked coequally and iteratively together during the planning phase. They made decisions together and committed to the results of the conceptualization process. Previous studies have also pointed out the social view of software ecosystems [e.g. 8]. Fricker [8] highlighted negotiation issues. Our case study indicates that negotiation issues are easier to avoid if the conceptualization process is done carefully together and the dependencies between tasks in the conceptualization process are clear and visible. The small number of actors also impacted on the ease of negotiation compared to large-scale software ecosystems with a correspondingly large number of actors.

Our paper presents how the business and social views were considered during the planning of the real-life software ecosystem. For practitioners, the paper provides knowledge on how actors can start building a software ecosystem together from a business perspective. In addition, our study points out the important role of RE activities to link the business view to the development of the MVP of digital services in the software ecosystem.

5.2 Threats to Validity

Here, we discuss four potential threats to the validity of the results. First, the interviews were conducted six months after the planning phase had ended. This might lead to deviations in the answers of the interviewees. This threat was mitigated by the researcher encouraging the interviewees to try to answer as they felt during the planning phase. In addition, the objectives of the study and the interviewee's rights and responsibilities were presented to them. The interviewees knew that the interviews were anonymous, and the material would be kept confidential. Therefore, it could be assumed that the interviewees gave honest answers.

Secondly, one of the limitations of this study is that only one representative from each actor was interviewed. Triangulation of the data sources was used to reduce this validity threat. The detailed material from the workshops was another source of data.

The third validity issue concerns investigator triangulation, which we were able to use in a restricted way. The first author of the paper was responsible for the design, execution, analysis and reporting of the study, and the second author reviewed the results of the study. The first author started to work at Case SECO after the planning phase, which enabled them to consider the planning phase neutrally. In addition, participation in Case SECO after the planning phase enabled her to understand the context and actors in detail.

The fourth limitation is that the findings of this study are derived from a single case study, where the case software ecosystem was quite small. It could be assumed that similar findings are achievable by conducting the same research, investigating the planning phase of another software ecosystem or repeating the same research for this case software ecosystem.

6 Conclusions

The results of this study give detailed information for practitioners on how to conceptualize digital services in the planning phase of a software ecosystem. The results show

how the business view can be incorporated systematically into the conceptualization of digital services and how actors can work together during the planning phase. If actors create the vision and objectives of the software ecosystem carefully together, it supports them in defining the MVP and prioritizing the functionalities of digital services. The results of the study also show that RE has a critical role in the planning phase of the software ecosystem. RE ensures traceability from the prioritized functionalities of the MVP to a value proposition, target customer groups and customer paths of digital services and further to the vision and objectives of the software ecosystem.

Our future research goal is to gain more detailed knowledge of how actors can conceptualize and develop digital services together in a software ecosystem. We also plan to conduct case studies and gather data from other software ecosystems in order to validate the findings of this study especially from the perspective of RE.

References

1. Bosch, J., Bosch-Sijtsema, P.: From integration to composition: on the impact of software product lines, global development and ecosystems. J. Syst. Softw. **83**(1), 67–76 (2010)
2. Boyce, C., Neale P.: Conducting in-depth interviews: a guide for designing and conducting in-depth interviews for evaluation input. In: Pathfinder International Tool Series, Monitoring and Evaluation, vol. 2 (2006)
3. Bukhsh, F.A., Bukhsh, Z.A., Daneva, M.: A systematic literature review on requirement prioritization techniques and their empirical evaluation. Comput. Stand. Interfaces **69**, 103389 (2020)
4. Campbell, P.R.J., Ahmed, F.: A three-dimensional view of software ecosystems. In: 4th European Conference on Software Architecture, pp. 81–84 (2010)
5. Carbone, P.: The emerging promise of business ecosystems. Technol. Innov. Manage. Rev. 11–16 (2009)
6. Charmaz, K.: Constructing Grounded Theory: A Practical Guide Through Qualitative Analysis, p. 208. Sage Publications (2006)
7. Dedehayir, O., Mäkinen, S., Ortt, R.: Roles during innovation ecosystem genesis: a literature review. Technol. Forecast. Soc. Change **136**, 18–29 (2018)
8. Fricker, S.: Specification and analysis of requirements negotiation strategy in software ecosystems. In: 1st International Workshop on Software Ecosystems, pp. 19–33 (2009)
9. Hanssen, G.K.A.: Longitudinal case study of an emerging software ecosystem: implications for practice and theory. J. Syst. Softw. **85**(7), 1455–1466 (2012)
10. Iansiti, M., Levien, R.: The Keystone Advantage: What the New Dynamics of Business Ecosystems Mean for Strategy, Innovation, and Sustainability. Harvard Business Press, Brighton (2004)
11. Jansen, S., Brinkkemper, S., Finkelstein, A.: Business network management as a survival strategy: a tale of two software ecosystems. In: 1st International Workshop on Software Ecosystems, pp. 34–48 (2009)
12. Jansen, S., Cusumano, M.A.: Defining software ecosystems: a survey of software platforms and business network governance. In: 4th International Workshop on Software Ecosystems, pp. 40–58 (2012)
13. Knodel, J., Manikas, K.: Towards a typification of software ecosystems. In: Fernandes, J.M., Machado, R.J., Wnuk, K. (eds.) ICSOB 2015. LNBIP, vol. 210, pp. 60–65. Springer, Cham (2015). https://doi.org/10.1007/978-3-319-19593-3_5

14. Lehtola, L., Kauppinen, M., Vähäniitty, J., Komssi, M.: Linking business and requirements engineering: is solution planning a missing activity in software product companies? Require. Eng. **14**(2), 113–128 (2009)
15. Manikas, K., Hansen, K.M.: Software ecosystems - a systematic literature review. J. Syst. Softw. **86**(5), 1294–1306 (2013)
16. Manikas, K.: Revisiting software ecosystems research: a longitudinal literature study. J. Syst. Softw. **117**, 84–103 (2016)
17. Markham, S.K., Ward, S.J., Aiman-Smith, L., Kingon, A.I.: The valley of death as context for role theory in product innovation. J. Prod. Innov. Manage. **27**(3), 402–417 (2010)
18. Moore, J.F.: The Death of Competition: Leadership and Strategy in the Age of Business Ecosystems. HarperBusiness (1996)
19. Pichlis, D., Raatikainen, M., Sevón, P., Hofemann, S., Myllärniemi, V., Komssi, M.: The challenges of joint solution planning: three software ecosystem cases. In: Jedlitschka, A., Kuvaja, P., Kuhrmann, M., Männistö, T., Münch, J., Raatikainen, M. (eds.) PROFES 2014. LNCS, vol. 8892, pp. 310–313. Springer, Cham (2014). https://doi.org/10.1007/978-3-319-13835-0_29
20. Rong, K., Shi, Y.: Business Ecosystems - Constructs, Configurations and the Nurturing Process. Palgrave Macmillan (2014)
21. Saarni, K., Kauppinen, M.: Activities and challenges in the planning phase of a software ecosystem. In: Hyrynsalmi, S., Suoranta, M., Nguyen-Duc, A., Tyrväinen, P., Abrahamsson, P. (eds.) ICSOB 2019. LNBIP, vol. 370, pp. 71–85. Springer, Cham (2019). https://doi.org/10.1007/978-3-030-33742-1_7
22. Schultis, K.-B., Elsner, C., Lohmann, D.: Architecture challenges for internal software ecosystems: a large-scale industry case study. In: 22nd International Symposium on Foundations of Software Engineering, pp. 542–552 (2014)
23. Valenca, G., Alves, C., Heimann, V., Jansen, S., Brinkkemper, S.: Competition and collaboration in requirements engineering: a case study of an emerging software ecosystem. In: 22nd International Requirements Engineering Conference, pp. 384–393 (2014)
24. Vegendla, A., Duc, A.N., Gao, S., Sindre, G.: A systematic mapping study on requirements engineering in software ecosystems. J. Inf. Technol. Res. **11**(1), 49–69 (2018)
25. Villela, K., Kedlaya, S., Doerr, J.: Requirements engineering for innovative software ecosystems: a research preview. In: Knauss, E., Goedicke, M. (eds.) REFSQ 2019. LNCS, vol. 11412, pp. 117–123. Springer, Cham (2019). https://doi.org/10.1007/978-3-030-15538-4_8
26. Yin, R.K.: Case Study Research: Design and Methods. Applied Social Research Methods, 3 edn. Sage Publications (2003)
27. Yu, E., Deng, S.: Understanding software ecosystems: a strategic modeling approach. In: 3rd International Workshop on Software Ecosystems, pp. 65–76 (2011)

Power and Privacy in Software Ecosystems: A Study on Data Breach Impact on Tech Giants

Maria Eduarda Rebelo, George Valença$^{(\boxtimes)}$, and Fernando Lins

Departamento de Computação, Universidade Federal Rural de Pernambuco,
Recife, Pernambuco, Brazil
{eduarda.rebelo,george.valenca,fernandoaires}@ufrpe.br

Abstract. [**Context and motivation**] Concerns about data privacy and protection in companies from various fields and sizes are not only a reality, but a requirement at this day and age. The need to comply with governmental laws and other rules became a driving force in handling personal data. [**Question/problem**] For major IT companies, especially those in charge of a software ecosystem, such concerns grow tenfold: cases of privacy breach can extend over and affect their platforms, software solutions and relationships with partners and users. [**Principal results**] This research investigates data breach cases in GAFA (Google, Amazon, Facebook, Apple) ecosystems through the perspective of power, which is a lens of analysis of a network of multiple interdependent actors. We create power models to describe the power relationships among ecosystem players during a privacy issue. [**Contribution**] Our descriptive case study reveals the actors involved in a data breach scandal, the ecosystem elements that grant them privileges or lack thereof, and consequences that reverberate positively or negatively towards them. We contribute towards stakeholder analysis activities by presenting our power relationships framework, which can be integrated into the requirements process as a technique for security and privacy requirements definition.

Keywords: Power · Privacy · Data protection · Ecosystem · Platform

1 Introduction

The consolidation of software ecosystems in the last decade represents a paradigm shift in the IT industry, in terms of both business models and software development. In this setting, varied actors adopt a platformisation approach to co-create value to a common pool of users, in a shared market for software and services. A keystone (e.g. Apple, Amazon) structures, releases, and controls a platform (e.g. iOS, Alexa), which is used by partners to complement or extend (e.g. creating or adapting a feature to a specific customer segment), or simply promote a software product (e.g. including it in an app store).

© Springer Nature Switzerland AG 2021
F. Dalpiaz and P. Spoletini (Eds.): REFSQ 2021, LNCS 12685, pp. 149–164, 2021.
https://doi.org/10.1007/978-3-030-73128-1_11

Ecosystems require software development to be oriented towards an architecture model that promotes secure software sourcing, integration, deployment, and evolution throughout a supply chain of different producers [17]. Otherwise, we may perceive events such as data breaches, which reveal the fragility of a software platform and related solutions. In 2017, a security researcher identified a data protection incident in the iOS ecosystem. The third-party solution AccuWeather from Apple's marketplace continued sending private location data to a backend monetisation service called RevealMobile, even with the location sharing turned off by the user [9]. A few years later, Amazon was accused of analysing snippets of conversations from Alexa-powered devices, which were secretly recorded and uploaded to the cloud without the user's consent [19]. In the Facebook ecosystem, a loose app review process combined with configuration errors allowed Instagram advertising partners to misappropriate a vast set of sensitive user data, including physical location and photos [16].

These examples highlight the need to ensure the privacy of user data, which is essential for software solutions to properly operate in the ecosystem. Furthermore, the success or failure of these solutions may affect the platform owner, innumerous complementors and, more importantly, the user base. Such impact is not perceived in isolation but rather in a systemic form, as the evolution of the ecosystem depends on the coopetition of its members. These actors share business, technical and social assets, e.g. the resources a keystone offers to the network, the expertise of a developer community; or the image of a respectable reseller, respectively [21]. Such assets are sources of power as they enable one party to increase another party's dependence by controlling what it values in the ecosystem [4]. Hence, power distribution becomes a useful lens of analysis of this network of multiple interfirm relationships and interdependent parties.

In this research, we investigate how actors in a software ecosystem exercise power in the occurrence of a data protection issue. To address this goal, we performed a descriptive case study of GAFA (Google, Amazon, Facebook, Apple) ecosystems, considering their relevance in the software industry and ubiquity in our daily routine. This study interpreted four critical privacy cases by creating power models to represent varied power relationships in these ecosystems. Such power framework can support stakeholder analysis activities in the early requirements phase, since understanding the social complexity around privacy breaches is paramount to define or assess data protection and security requirements [10], which is especially critical in complex settings like software ecosystems [24].

The rest of the paper is structured as follows. Section 2 describes the conceptual framework of this research, over which we present our findings. Section 3 describes our methodology. In Sect. 4, we detail privacy breach cases in GAFA ecosystems in terms of power relationships. Finally, in Sect. 5 we discuss the implications or our contribution, compare our results with other works, and present threats to validity as well as future work.

2   Theoretical Foundation

2.1   Software Ecosystems

Software ecosystems can be described as a set of businesses functioning collectively as a unit and interacting with a shared market for software and services, forging relationships among themselves. They gather actors investing in innovative business models, who aim to co-create value for the ecosystem. In a platform business model, companies open their platforms for third parties/potential partners to integrate their specific solutions and/or develop new ones, in a movement from single to multiple products in a platform approach [15]. Well-known examples of ecosystems include Microsoft's Dynamics CRM and Apple's iOS.

It is possible to comprehend how software ecosystems operate by describing its three different **dimensions** [21]. The *social* dimension encompasses the actors participating in the ecosystem with their respective roles, relationships, skills and motivations, among other factors that regulate the interactions within the network. The *technical* dimension is primarily concerned with the software platform itself and its software-based system that provides core features shared by a portfolio of products or services that interoperate with each other, with extended solutions via boundary resources, such as application programming interfaces (APIs) and software development toolkits (SDKs) [15]. Within this dimension, product management and development processes that shape how solutions are collaboratively planned, evolved and released to customers can also be found. Lastly, the *business* dimension deals with the strategies to obtain value and generate revenue for all ecosystem participants by involving the platform business model and its definitions about entry barriers and intellectual property rights, as well as overall innovation directions [21].

Broadly speaking, three main groups of **actors** form an ecosystem. Those in charge of controlling and those subsumed to the current rules. A firm called *keystone* leads the evolution of the ecosystem by defining rules of access to a platform and orchestrating the creation of new solutions. In parallel, several *complementors* co-create value on top of the platform by combining their solutions to address the needs for features/services from a wide pool of *users* [13].

Ecosystems emerge from a software platform whose architecture enable an ease extension of its functionality via APIs, for instance. However, such platform may face serious security risks, such as malicious code spread from solutions developed by ecosystem partners, which may expose user data or ultimately disable the overall system. Hence, **security and privacy requirements** are relevant parameters to design and evaluate the architecture of a platform [24].

Data protection laws such as the EU's General Data Protection Regulation (GDPR) and the California Consumer Privacy Act recently introduced a hard challenge for ecosystems. Keystone companies such as Google and Amazon must ensure that their platform and derived software solutions from niche players meet the requirements at the core of such regulations to protect users' personal data and privacy. Any deviance from the prescribed rules can be easily noticed due to the large customer base around these ecosystems.

2.2 Power

Social scientists have studied power in interpersonal relationships for decades. Behavioral scientists and management professionals later introduced this concept in a business environment to analyse interfirm relationships, when a company can hold and exercise power while interacting with another company. Based on a previous theoretical interpretation of power in software ecosystems [21], we define the power of A as its ability to exert some sort of influence in its relationship with B. The resulting power relationships are often based on an underlying idea of dependence, e.g. an actor A can exercise power over an actor B if this player somehow depends on A [4]. Such power may occur in five different **forms**, according to the well-known taxonomy from French and Raven [6]:

- *Coercive* power is B's perception that A can punish it (e.g. a keystone disqualifies partners whose products do not live up to quality standards).
- *Reward* power is B's perception that A can provide it with rewards (e.g. a company offers financial benefits to ecosystem partners).
- *Expert* power is B's perception that A has special knowledge or expertise (e.g. a company masters innovative technologies).
- *Legitimate* power is B's perception that A has the right to impose behavior for it (e.g. a company can set ecosystem goals due to its superior position).
- *Referent* power is B's feeling of respect, admiration and identification toward A (e.g. partners recognise the status of the keystone).

A **power capability** is a given asset that denotes a company's power, such as developing features for a specific market segment, providing partners with key information about customers, or establishing the roles of partners in the ecosystem. Each power capability derives from power sources, which are tangible or intangible resources that an actor can use to affect the behaviour of others. Thereby, a company can exercise some form of power by cultivating such sources. Any change in the availability or demand for power sources may affect the power distribution in a partnership, since it causes an player to gain or lose power [21].

3 Research Method

This research reports on a descriptive case study of privacy breaches in the well-established GAFA software ecosystems. We considered that all the actors involved in such scandals hold a certain level of power, which reflects upon countless business and technical decisions on the companies' end in an attempt to repair the damage caused to the users. By combining web and literature-based information, we analysed how privacy issues affected power relationships established around influential players. Figure 1 presents the main phases of the study (data collection, analysis and synthesis), which we describe as follows.

Our **data collection** aimed at mapping articles reporting data leak cases in software ecosystems. Accordingly, we defined and calibrated a search string, whose final structure involved the notions of privacy, breach and ecosystems

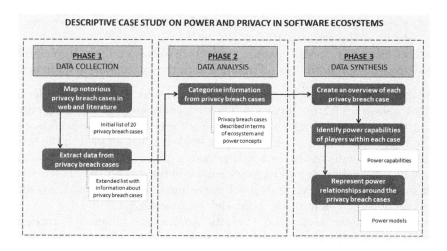

Fig. 1. Phases and respective activities of our descriptive case study.

(in the scope of GAFA): *privacy AND (scandal OR breach OR data leak) AND (Google OR Amazon OR Facebook OR Facebook)*. This query was applied in Google's search engine considering a timeframe of five years[1]. To guarantee the quality of web-based data, our selection criteria was based on two indicators: the "Media Bias Rating" scale[2], which rates news and editorial/opinion content for a number of outlets, and the "Fact Check"[3] list, which rates news media sources based on factual accuracy and political bias. Hence, we could assess the reputation of the sources behind each article, which restricted our results to relevant IT news websites and large-circulation outlets (e.g. Forbes, The Guardian).

Then, we performed a literature review to retrieve additional data about privacy breaches in ecosystems. This step provided us with works such as [9], which reports privacy issues in Facebook and Apple/iOS ecosystems. In total, we selected 20 articles across 14 different web and literature publications for further analysis. Altogether, we identified 5 cases per ecosystem within the total of articles, which included peer-reviewed papers and newspapers texts.

To structure our dataset, we extracted the following information (which we illustrate with an example from iOS ecosystem) about the selected breach cases:

- *Publication reporting the case* (e.g. article - Forbes; or event - 52nd HICSS)
- *Data breach description* (e.g. app AccuWeather disclosing user device data with backend services such as RevealMobile in iOS platform)
- *Internal/external actors involved in the case* (e.g. Apple, AccuWeather)
- *Products and platform* (e.g. AccuWeather, iOS)
- *Actor who caused the data breach* (e.g. AccuWeather)

---

[1] We date back to 2016 since that year is the start of a time period encompassing multiple security breach scandals, such as the Facebook-Cambridge Analytica.

[2] Media Bias Ratings - https://www.allsides.com/media-bias/media-bias-ratings.

[3] Media Bias/Fact Check - https://mediabiasfactcheck.com.

- *Relevance and impact of the data breach* (e.g. a virulence was present and brought to public attention)
- *Actions taken* (e.g. media pressure made AccuWeather remove RevealMobile SDK from its app until it fully complied with the appropriate requirements).

We extracted data from these cases in a spreadsheet that acted as a form of breakdown sheet for further analysis[4]. Then, we considered the following criteria to prioritise and select a critical case from each ecosystem: (i) relevance of the scandal (e.g. social and media repercussion, evidence that alluded to the breach having negative implications on the keystone at play); (ii) level of detail (e.g. amount of information regarding post-scandal actions taken by the company, which would allow a proper analysis of the elements involved); and (iii) direct quotations of responses from the keystone's representatives.

In **data analysis**, we adopted Thematic Analysis guidelines [3] to categorise the evidence related to the four cases. This coding process allowed us to describe the scenario around each case by using themes related to software ecosystem and power. Initially, we mapped key elements from each case (actors, affected software product or platform, consequences) to software ecosystem nomenclature [21]. For instance, an article excerpt such as *"regulators said that YouTube, which is owned by Google, had illegally gathered children's data"* provided us with information about the actors (keystone - Google; users - children; third party/external entity - legal representatives) and software platform/product (Youtube). Then, we identified and interpreted evidence of power exercise within cases description based on the substantive theory of power and dependence in ecosystem partnerships [21], which presents an approach for examining the exercise of power in ecosystems together with illustrative power models. For instance, in the excerpt *"users will be able to opt-in to help Siri improve by learning from the audio samples of their requests"*, we identified a right or prerogative from users in the case when Apple was caught sending Siri recordings to contractors worldwide. Hence, we labeled it as an occurrence of legitimate power.

Finally, the **data synthesis** phase was dedicated to thoroughly interpret the power relationships among ecosystem actors. Our goal was twofold: (i) create a comprehensive narrative of the data breach event; and (ii) describe what and how power capabilities were used by ecosystem players in each event[5]. After creating an overview of each case, we identified evidence of power exercise and associated power sources. For example, in Google case, we perceived that attorneys and trade commission officers penalised Google financially and demanded privacy changes under legal settlement. We understood this fact as a coercive power capability of legal entities to notify legal authorities about YouTube's law violations on software product, whose source lied in the legal permissions to investigate disobedience through knowledge of established laws. By relating power capabilities to each involved actor, we represented their power relationships in power models (e.g. such coercive power is represented as a box related to the actor "legal entities" with an arrow directed to the actor "keystone", over

---

[4] Analysis spreadsheet - https://bit.ly/3pHMx13.

[5] Detailed power capabilities of actors involved in the cases - https://bit.ly/35CEydA.

who power is exercised). Besides, we examined the actions taken once the breach was exposed, considering aspects such as privacy policy changes, redefinition of products' requirements, and financially prohibitive measures.

4 Results

4.1 Case 1 – YouTube (Google Ecosystem)

In a case reported by New York Times, YouTube (which is owned by Google) illegally gathered children's data (including identification codes used to track web browsing over time) without their parents' consent [18]. According to the accusations, Youtube marketed itself to advertisers as a top destination for young children, despite informing these firms that they did not have to comply with the children's privacy law because YouTube did not have viewers under the age of thirteen. Youtube then proceeded to make millions of dollars by using the information harvested from these children to redirect specific ads their way.

This investigation caused Google to repair the damage by paying a record $170 million fine and to make needed changes to secure children's privacy on their YouTube platform. Such move resulted from an enforcement action taken by U.S. regulators against technology companies for violating users' privacy. Claims suggested that Youtube had knowingly and illegally harvested personal data from children and used it to profit by targeting them with ads. As part of the settlement, YouTube also agreed to create a system that asks video channel owners to identify the children's content they post so that targeted ads are not placed in such videos. The regulators stipulated that YouTube must also obtain consent from parents before dealing with personal details like a child's name or photos. Despite the significant settlement, regulators and other legal entities were critical of how the case was conducted (a U.S. Senator described the penalty as *"a slap on the wrist for one of the world's richest companies"* [18]).

For the past few years, YouTube has redirected its efforts to accommodate underage users on its platform and service. The exercise of **expert power** can be identified in how YouTube has orchestrated its marketing strategies to make advertisers recognise its platform as a top destination for young children, despite telling them it would not require any compliance with children's privacy laws as YouTube *"did not have users under 13"* [PC_EXYT01] [18]. The expert power exercised by YouTube is such that users are not necessarily suspicious of any malicious activity, with factors such as reputation and trust coming to play in the relationship between the keystone and customer base. This comfort zone generated by the exercise of expert power granted YouTube a silent permission to illegally gather, monitor, and track children's data without their parents' consent, as well as to serve targeted ads to young children [18].

New York's Attorney General Letitia James, responsible for enforcing the federal children's privacy law in the state, notified the trade commission of apparent violations of the law on the site [18], which resulted in the penalty and changes encompassed in the settlement along with the Federal Trade Commission [PC_COYT01]. The accusations against YouTube pointed to a direct

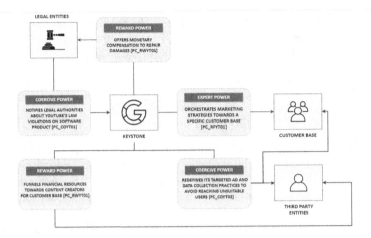

Fig. 2. Power model for YouTube/Google case.

violation of the federal Children's Online Privacy Protection Act. Google sought to make amends for requirements they should have been complying with in the first place. It began by offering financial compensation to repair damage under the legal settlement, reinforcing its ability to exercise **reward power** [PC_RWYT01].

The scandal cornered YouTube into taking firmer and visible actions that reach beyond the users alone, proving its ability to exercise **coercive power** upon third-party entities involved. It agreed to not only stop placing targeted ads on children's videos, but also prevented complementors from gathering personal data about anyone who watched such videos, even if the viewer could be an adult [18]. Hence, Youtube limited how much video makers earn on the platform, as they will no longer be able to profit from ads targeted at children [PC_COYT02].

Another action on YouTube's end can be understood as **reward power** upon users and third-party entities involved. These actions involved funneling $100 million to creators of children's content over the next three years after 2019. [18] [PC_RFYT01]. Additionally, YouTube claimed it would *"heavily promote YouTube Kids, its child-focused app, to shift parents away from using the main YouTube app when allowing their children to watch videos"* [18]. In Fig. 2, we represent the use of such power capabilities by actors involved in this case.

4.2 Case 2 – Alexa (Amazon Ecosystem)

To illustrate this particular case, three different reports regarding the same subject matter were concatenated for further analysis. The core of the these privacy breach events from Amazon concern Alexa, the popular and well-established voice assistant built in certain Amazon devices, such as the Echo speakers. Both Amazon Echo and Alexa voice assistant have had widely publicised issues with privacy [2] and it branches into many problematic topics that have been exposed by the media on multiple occasions. Amazon attributed the error to Echo mishearing the "wake" word, which led to a request to send a message; then, mishearing a name in contacts list and a confirmation to send the message. By analysing Alexa's transcripts, Bloomberg identified that it did wake up accidentally in more than one out of 10 transcripts. These dangerous slips on Amazon's end raised major concerns on the way Alexa devices interact with other services, directly *"risking a dystopian spiral of increasing surveillance and control"* [2].

An article from the Washington Post highlighted problems related to Alexa's data capture with a report on users being unable to take any actions to prevent Amazon from collecting data other than muting the device's microphone altogether [5]. This issue is also linked to another practice implemented by the company where recordings are listened and reviewed by human contractors under the argument of *"[listening] to recordings to train its artificial intelligence"* and admittedly reported that *"some of those employees also have access to location information for the devices that made the recordings"* [5].

The most concerning factor is that Amazon is disturbingly quiet, evasive, and reluctant to act when it comes to tackling the privacy implications of their practices, many of which are buried deep within their terms and conditions or hard-to-find settings [2]. Whether it is the amount of data Amazon collects or the fact it reportedly pay employees (and, at times, external contractors globally to listen to recordings to improve accuracy), the potential exists for sensitive personal information to be leaked through these devices. Criticism towards the way this player handles personal data and its practices is also questioned, accompanied by actions that should be required in order to respect these boundaries.

Amazon's biggest advantage lies in the exercise of **expert power** over its customers. By *"accurately interpreting voice commands by taking account of different languages, accents, tones, contexts and degrees of ambient clutter"* [11], Amazon improves its services. As these solutions become practical, they bring a sense of trust and convenience that retain users in the ecosystem [PC_RFAZ01].

Such expert power is nurtured by Amazon's **legitimate power** of analysing user data to provide services that are painstakingly designed for the customers. Amazon uses this legitimate power by exploring its knowledge of what users are searching for, listening to, or sending in personal messages [PC_LGAZ01]. This gives Amazon a large control over the customer base's data [2]. However, there are concerns about such power capability as it is enabled by Amazon's invasive and even obscure practices, since users are not aware of the extent and nature of the data harvesting. For instance, Amazon signed a deal with UK's National Health Service for medical advice provided by the Echo assistant, which could lead to users' health data getting linked to online shopping suggestions and even third-party ads causing some sort of oversharing with the company [2].

It is also possible to identify its **legitimate power** of using third-party services as a pool of data from which Amazon services and products can collect information [PC_LGAZ02] – raising concerns regarding how Alexa devices interact with other services [2]. Nonetheless, Amazon acknowledges it collects data about third-party devices even when users do not utilise Alexa to operate them.

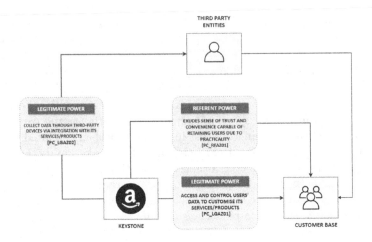

Fig. 3. Power model for Alexa/Amazon case.

It also mentions Alexa needs to know the "state" of users' devices *"to enable a great smart home experience"* [5], while it is fairly unlikely that customers are aware of this practice among other powerful ecosystems as a whole. Figure 3 shows the use of such power capabilities in light of the data protection issue described.

4.3   Case 3 – Instagram (Facebook Ecosystem)

In August 2019, Business Insider reported on a combination of configuration errors and lax oversight by Instagram that allowed Hyp3r, a vetted advertising partner from the social network, to misappropriate vast amounts of public user data. Hyp3r created detailed records of users' physical whereabouts, personal bios, and photos that were intended to vanish after 24 h [16]. This partner developed a tool to "geofence" specific locations and then harvest every public post tagged with that location on Instagram.

Hyp3r scraped and stitched users' profiles together, which constituted a clear violation of Instagram's rules. However, it all occurred on Instagram's watch throughout 2019. In particular, Hyp3r was considered by Instagram as one of its preferred "Facebook Marketing Partners" [16]. Stories (supposed to disappear after 24 h) from ordinary users of Instagram have never been available through Instagram's API. Hyp3r orchestrated a way to also collect this type of data, saving the temporary images indefinitely, despite Instagram's requirement to store content only "for the period necessary to provide your app's service".

The unauthorised use of Instagram data by Hyp3r involved *"[taking] advantage of an Instagram security lapse, allowing it to zero in on specific locations, like hotels and gyms"*. Besides, it collected user bios and followers, which were then combined other sources, such as user location [16].

Hyp3r also neglected the prohibition on "reverse engineer[ing] the Instagram's APIs". It deliberately rebuilt its own version of an API that Instagram shuttered after Cambridge Analytica [16]. The result included a database of thousands of locations. Additional information also revealed a publicly available JSON package that bundles up various bits of data in an easy-to-access format, when users access Instagram through their web browsers. No logging in is required to gain approval or to authenticate one's identity in any way to access it, which denotes an unexpected breach on Instagram's end [16].

The issue regarding Instagram and Hyp3r demonstrates one of Facebook's biggest struggles when it comes to restricting users' personal information and the way it extends beyond the core of their main Facebook app. Instagram is certainly the only service affected over the years, but Hyp3r is probably not the only business scraping its data [16]. Hence, Hyp3r's activity raises questions about the extent of the due diligence that Facebook conducts on partners using its ecosystem, as well as on its procedures to safeguard user data.

Despite these facts, Hyp3r denied breaking Instagram's requirements. This partner argued that it accesses public data on Instagram. The result of the public information it gleaned was a sophisticated database about Instagram users' interests and movements. Hyp3r openly touted such database to customers as one of its key selling points, despite the fact that Instagram's policies were structured so that such a thing would not be possible.

Instagram responded the scandal by sending Hyp3r a cease-and-desist letter confirming this ecosystem partner broke the rules of the social network [16]. Hyp3r then claimed to process public data, whose harvest does not require consent from Instagram users. It also added that companies have legitimate business needs that justify knowing what is being shared from their properties [16].

Naturally, an exercise of **expert power** is immediately identified from Instagram over Hyp3r, since the keystone is a source of crucial information for the advertiser solution to become what it is [PC_EXFB01]. Through such relevant user data, Hyp3r orchestrated its marketing strategies to attract customers [16].

Despite the controversy surrounding Hyp3r's practices to successfully collect data from Instagram, this partner reinforced such database was accessed through legit means. Hence, Hyp3r was exercising **legitimate power** over Instagram with a retort that argues how *"accessing public data on Instagram in this way is legitimate and justifiable"* [16] [PC_LGFB01]. Through the partnership with Facebook ecosystem, Hyp3r can also exercise **reward power** over its customers. Its groundbreaking services rely heavily on exploring data from Instagram users to package and sell a customised marketing strategy to customers [PC_RWFB01].

Hyp3r's claimed to follow the rules from Instagram in the partnership with Facebook. However, its data scraping appeared to violate multiple requirements, such as "to store or cache content only for as long as necessary to provide a required service" [16]. Hence, the company stored user data indefinitely. By prohibiting the practice of the so-called "reverse engineer[ing]" of Instagram's APIs, Facebook's exercise of **coercive power** over Hyp3r (a power it maintains over its partners in general). The keystone stepped in and *"completely revoked Hyp3r's access to its APIs, removed it from the list of Facebook Marketing Partners"* [16], *despite initially including Hyp3r on an exclusive list of partners* [PC_COFB01], [PC_COFB02]. Additionally, an exercise of **coercive power** of media outlets

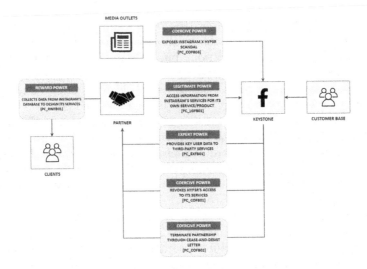

Fig. 4. Power model for Instagram/Facebook case.

over Facebook is identified through the published report of this scandal, which eventually led to Instagram sending Hyp3r *"a cease-and-desist letter after being presented with Business Insider's findings and confirmed that the startup broke its rules"* [16], illustrating a chain reaction of power exercising [PC_COFB03]. The use of such power capabilities by ecosystem actors is presented in Fig. 4.

4.4 Case 4 – Siri (Apple Ecosystem)

For this case, we considered two complementary articles published by Forbes. They illustrate the scandal scenario concerning Apple's voice assistant Siri, whose actors and power capabilities we represent in Fig. 5 and describe as follows. These reports highlight that *"a small proportion of Siri recordings are passed on to contractors working for the company around the world"* [20]. This practice was described as a means to improve Siri after its accidental activation, either through [Apple's] smartwatch, the HomePod wireless speaker or one of the other Apple mobile devices including the iPhone, the iPad, or the iPod touch [20]. This situation was labeled as a concerning privacy gaffe and raised questions regarding Apple's practices when it comes to handling private data from customers. Moreover, customers may doubt whether or not Apple practices what it preaches: there is *"a false sense of privacy that Apple has communicated through its marketing strategy to distinguish itself from Amazon and Google"* [20].

Another report from the Guardian revealed that Apple contractors were regularly hearing confidential details on customers' Siri recordings[12]. This evidence led the company to (i) promptly review its process of handling the recordings of Siri queries, and (ii) announce it would turn off recordings by default and bring the human evaluation process in-house [19]. In particular, Apple clearly stipulates in its privacy policy that it does send to its servers data such as *"your name, contacts [...] and searches to help Siri [...] provide better responses"* [20].

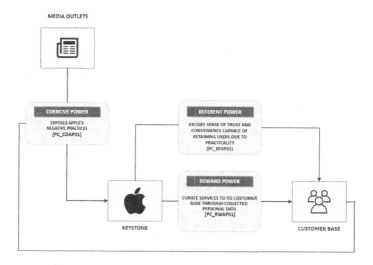

Fig. 5. Power model for Siri/Apple case.

Apple is known for its strong reputation, which derives from the kind of experience it offers and the convenience of using its integrated services. This reinforces its **referent power** over its customers. However, such power capability causes the mishandling of data to be rarely considered until a critical data breach comes to light [PC_RFAP01]. Apple seems to not only collect an overwhelming amount of data, but also transfer these scraps of data, which are *"carefully assembled, synthesized, traded, and sold"* [20] to third-party solutions in its ecosystem. Unlike other key players such as Amazon (Alexa) and Google (Google Assistant), Apple does not provide means to opt-out having audio recordings sent to servers.

However, it is possible to identify the exercise of **coercive power** by media outlets and the press (and, consequently, users once they get to read these articles) as entities capable of exposing these negative practices. Once these publications revealed the case, Apple decided to thoroughly review the process that it uses to handle the recordings of Siri queries [19] [PC_COAP01]. This situation caused Apple to announce that it would turn off recordings by default, as well as *"bring the human evaluation process in-house"* [19].

The core of the problem lies in Apple privacy policy stipulating that certain personal information is sent to its servers (e.g. users' names and list of contacts to enable Siri to provide precise responses, enhancing this software solution). This creates *"a false sense of privacy with their marketing messaging"* [20], which paves the way for Apple to exercise **reward power** over the customer base: as users overshare their data, Apple offers better curated personal service. The tech giant argues such data is needed because *"[the] goal with Siri, the pioneering intelligent assistant, is to provide the best experience for our customers while vigilantly protecting their privacy"* [19] [PC_RWAP01].

5 Discussion and Conclusion

5.1 Implications for Research and Practice

We investigated data protection issues in the widespread software ecosystem setting through the lenses of power. Since security and privacy issues stem from intentions and concerns of actors (e.g. company, users, suppliers) [10], we created power models that illustrate the ecosystem players and interactions during events of data breach and threats to user privacy. Our analysis of power exercise is related to the core area of stakeholder analysis in the requirements engineering process. We believe business analysts can benefit from identifying power capabilities and their respective use by actors forming the operational environment of ecosystem solutions. Similarly to other actor-oriented techniques such as i*, the resulting power models can act as a decision support technique at the early requirements stage to represent how actors behave and influence each other [7].

Based on the power models, we can start mapping conflicts and synergies among ecosystem actors, as well as implications of privacy issues for their relationships. Such blueprint of the context around the platform enables a team to further derive quality and functional requirements for data protection. For instance, through the power model in Fig. 2, one can perceive Google using coercive power to prevent complementors from gathering personal data about anyone who watched videos target at children. To avoid the use of such negative form of power, which weakens partnerships and the migration of third parties [22], the keystone could arrive at business rules such as "ensuring that only data genuinely needed are collected by complementors" or "ensuring that complementors only get access to data they are (legally) entitled to" [23].

Concerns with data and privacy protection are growing in the software industry, and more so regarding big influential tech companies such as the ones leading the relevant software ecosystems we described. With the implementation of data protection regulations and the need to consider privacy by design, we suggest the analysis of power relationships at the earliest stages of requirements engineering. The current power models can be evolved by the requirements community to represent the elements of the ecosystem used as sources of power.

5.2 Threats to Validity

Information regarding the privacy breaches were primarily collected through the articles reported by selected publications along with their presented evidence. Despite the careful analysis of our sources (via Fact Check tool) and regardless of how informative the reports were, the selected journalistic pieces were susceptible to personal impressions of the publication author's, which may have have affected the *internal validity* of this study. Hence, additional undisclosed factors could also influence the analysis. Our findings derived from a qualitative process, with the analysis of subjective information. To avoid threats to *conclusion validity*, i.e. an inappropriate interpretation of the scenarios, we selected cases with a great amount of information, which we concentrated in one dataset. Besides, during three cycles, the researchers analysed the evidence and discussed its classification to refine the interpretation of privacy and power in GAFA ecosystems.

5.3   Related Work

The work from Milne and Maiden [14] also adopted French and Raven's power taxonomy [6] to explore the role of power and politics, which they consider critical factors in the requirements engineering process. Our representation of power relationships among ecosystem actors (interfirm interactions) was inspired by their structure of power forces in social networks (social interactions). In their turn, Hurni and Huber (2014) [8] studied the interplay of power and trust on relationships among actors in big platform ecosystems, but did not approach requirements nor data protection issues.

5.4   Future Work

In terms of research method, we plan to collect complementary data about data protection issues in ecosystems. By conducting interviews with relevant stakeholders from each of these ecosystems (e.g. developers, users) and considering other media outlets (e.g. videos) reporting information on the cases and overall situation surrounding the scandals, we will be able to clarify and enhance the veracity of the facts. The studied and additional cases can also enable the identification of recurrent patterns of power exercise during data breach scandals.

To enhance our contribution, we aim to specify a power relationships analysis within privacy-oriented processes such as GuideMe [1], a systematic approach to elicit solution requirements in light of data protection regulations and stakeholder scenario. Hence, we can better situate our power framework as a technique for RE and ecosystem practitioners to perform a contextual analysis of privacy. Moreover, we aim to define privacy requirements for ecosystems by interpreting data protection laws such as GDPR, based on privacy challenges we previously mapped [23]. Such requirements can guide keystones towards compliance with data protection laws and prevent other breach scandals.

References

1. Ayala-Rivera, V., Pasquale, L.: The grace period has ended: an approach to operationalize GDPR requirements. In: 26th International RE Conference, pp. 136–146 (2018)
2. Benjamin, G.: Amazon echo's privacy issues go way beyond voice recordings, January 2020. https://theconversation.com/amazon-echos-privacy-issues-go-way-beyond-voice-recordings-130016. Accessed 29 July 2020
3. Cruzes, D.S., Dyba, T.: Recommended steps for thematic synthesis in software engineering. In: 5th ESEM, pp. 275–284. IEEE (2011)
4. Emerson, R.M.: Power-dependence relations. Am. Sociol. Rev. **27**(1), 31–41 (1962). http://www.jstor.org/stable/2089716
5. Fowler, G.: Alexa has been eavesdropping on you this whole time, May 2019. https://www.washingtonpost.com/technology/2019/05/06/alexa-has-been-eavesdropping-you-this-whole-time/. Accessed 02 Aug 2020
6. French, J., Raven, B.: The bases of social power, vol. 6, January 1959

7. Horkoff, J., Yu, E.: Interactive goal model analysis for early requirements engineering. Requirements Eng. **21**(1), 29–61 (2014). https://doi.org/10.1007/s00766-014-0209-8
8. Hurni, T., Huber, T.: The interplay of power and trust in platform ecosystems of the enterprise application software industry (2014)
9. Kurtz, C., Wittner, F., Semmann, M., Schulz, W., Böhmann, T.: The unlikely siblings in the GDPR family: a techno-legal analysis of major platforms in the diffusion of personal data in service ecosystems, January 2019
10. Liu, L., Yu, E., Mylopoulos, J.: Security and privacy requirements analysis within a social setting. In: 11th IEEE International Requirements Engineering Conference, pp. 151–161 (2003)
11. Lynskey, D.: Alexa, are you invading my privacy?, October 2019. https://www.theguardian.com/technology/2019/oct/09/alexa-are-you-invading-my-privacy-the-dark-side-of-our-voice-assistants. Accessed 29 July 2020
12. Lynskey, D.: Apple contractors regularly hear confidential details on Siri recordings, July 2019. https://www.theguardian.com/technology/2019/jul/26/apple-contractors-regularly-hear-confidential-details-on-siri-recordings. Accessed 29 July 2020
13. Manikas, K., Hansen, K.M.: Software ecosystems – a systematic literature review. J. Syst. Softw. **86**(5), 1294–1306 (2013)
14. Milne, A., Maiden, N.: Power and politics in requirements engineering: a proposed research agenda. In: 19th RE Conference, pp. 187–196. IEEE (2011)
15. Nambisan, S., Siegel, D., Kenney, M.: On open innovation, platforms, and entrepreneurship. Strateg. Entrep. J. **12**(3), 354–368 (2018)
16. Price, R.: Instagram's lax privacy practices let a trusted partner track millions of users' physical locations, secretly save their stories, and flout its rules, August 2019. https://www.businessinsider.com/startup-hyp3r-saving-instagram-users-stories-tracking-locations-2019-8. Accessed 29 July 2020
17. Scacchi, W., Alspaugh, T.A.: Securing software ecosystem architectures: challenges and opportunities. IEEE Softw. **36**(3), 33–38 (2018)
18. Singer, N., Conger, K.: Google is fined $170 million for violating children's privacy on youtube, September 2019. https://www.nytimes.com/2019/09/04/technology/google-youtube-fine-ftc.html. Accessed 10 Aug 2020
19. Su, J.: Apple apologizes for eavesdropping on customers, August 2019. https://www.forbes.com/sites/jeanbaptiste/2019/08/28/apple-apologizes-for-eavesdropping-on-customers-keeping-siri-recordings-without-permission/. Accessed 05 Aug 2020
20. Su, J.: Confirmed: apple caught in Siri privacy scandal, July 2019. https://www.forbes.com/sites/jeanbaptiste/2019/07/30/confirmed-apple-caught-in-siri-privacy-scandal-let-contractors-listen-to-private-voice-recordings/. Accessed 25 July 2020
21. Valença, G., Alves, C.: A theory of power in emerging software ecosystems formed by small-to-medium enterprises. J. Syst. Softw. **134**, 76–104 (2017)
22. Valença, G., Alves, C., Jansen, S.: Strategies for managing power relationships in software ecosystems. J. Syst. Softw. **144**, 478–500 (2018)
23. Valença, G., Kneuper, R., Rebelo, M.E.: Privacy in software ecosystems-an initial analysis of data protection roles and challenges. In: 46th Euromicro Conference on Software Engineering and Advanced Applications, pp. 120–123 (2020)
24. Vegendla, A., Duc, A.N., Gao, S., Sindre, G.: A systematic mapping study on requirements engineering in software ecosystems. J. IT Res. **11**, 49–69 (2018)

Iterative and Scenario-Based Requirements Specification in a System of Systems Context

Carsten Wiecher[1]([✉])[ID], Joel Greenyer[2][ID], Carsten Wolff[1][ID], Harald Anacker[3], and Roman Dumitrescu[3]

[1] Dortmund University of Applied Sciences and Arts, 44139 Dortmund, Germany
{carsten.wiecher,carsten.wolff}@fh-dortmund.de
[2] FHDW Hannover, 30173 Hannover, Germany
joel.greenyer@fhdw.de
[3] Fraunhofer IEM, 33102 Paderborn, Germany
{harald.anacker,roman.dumitrescu}@iem.fraunhofer.de

Abstract. [Context & Motivation] Due to the managerial, operational and evolutionary independence of constituent systems (CSs) in a System of Systems (SoS) context, top-down and linear requirements engineering (RE) approaches are insufficient. RE techniques for SoS must support iterating, changing, synchronizing, and communicating requirements across different abstraction and hierarchy levels as well as scopes of responsibility. [Question/Problem] We address the challenge of SoS requirements specification, where requirements can describe the SoS behavior, but also the behavior of CSs that are developed independently. [Principal Ideas] To support the requirements specification in an SoS environment, we propose a scenario-based and iterative specification technique. This allows requirements engineers to continuously model and jointly execute and test the system behavior for the SoS and the CS in order to detect contradictions in the requirement specifications at an early stage. [Contribution] In this paper, we describe an extension for the scenario-modeling language for Kotlin (SMLK) to continuously and formally model requirements on SoS and CS level. To support the iterative requirements specification and modeling we combine SMLK with agile development techniques. We demonstrate the applicability of our approach with the help of an example from the field of e-mobility.

Keywords: System of systems engineering · Requirements analysis · Requirements specification · Scenario-based requirements modeling

1 Introduction

New methods and tools are needed to meet the challenges in the development of complex socio-technical systems, such as sustainable mobility solutions in metropolitan regions [25]. Systems of connected electrified vehicles can be characterised as a *system of systems* (SoS), where the vehicle can be seen as a

© Springer Nature Switzerland AG 2021
F. Dalpiaz and P. Spoletini (Eds.): REFSQ 2021, LNCS 12685, pp. 165–181, 2021.
https://doi.org/10.1007/978-3-030-73128-1_12

constituent system (CS) that interacts with changing other CSs to provide an SoS functionality [17].

An interdisciplinary approach for the realization of these systems is *system of systems engineering* (SoSE). The definition of stakeholder needs and required functionalities are key elements of SoSE [20]; the precise specification of requirements is a basis for the system decomposition and implementation, or the selection of suitable CSs that form an SoS [27]. However, in SoSE, there are different requirements engineering (RE) challenges compared to RE in established systems engineering (SE) processes [26].

According to Maier et al. [23], the operational, managerial, and evolutionary independence are the essential characteristics of an SoS. These characteristics have a significant influence on the applicability of existing RE techniques [24–26]. In contrast to monolithic systems, SoS consist of individual systems that can operate independently and perform a meaningful task, even when not part of an SoS. The development and operation of the CSs is managed independently, in different organizations with different development- and product life cycles. Also, requirements on the CS- and SoS level change frequently and independently, leading to an evolutionary development [23, 26].

Based on these SoS characteristics, Ncube and Lim [24] describe challenges for the SoS RE process: Due to the different systems in an SoS, requirements cover many different disciplines, can be contradictory, unknown or possibly not fully defined. These difficulties overlap with the fundamental problems in RE [7], but, according to Ncube and Lim [24], requirements in an SoS additionally must be considered as requirements for the SoS, which describe the properties of the overall system, or requirements for a CS that describe capabilities of a single system. Since requirements on both levels can change continuously and independently, traditional, linear and top-down requirements specification and decomposition techniques can not be used [24–26].

To address this problem we propose an iterative and scenario-based requirements specification technique. Based on previous work [30, 32, 33] we integrate the Scenario Modeling Language for Kotlin (SMLK) with agile development techniques to support the requirements engineer in the continuous and iterative specification, formalization, and validation of requirements on different levels of abstraction.

This paper makes the following two contributions: First (1), we extend SMLK to enable requirements engineers to intuitively, but formally model the requirements on the SoS-level as well as the interaction between the CSs (CS-level). With these extensions, requirements can be specified and validated independently, which addresses the managerial and operational independence of systems. Nevertheless, both levels of abstraction are connected to allow for the joint execution and testing of the specified behavior on the SoS- and CS-level, in order to detect and resolve contradictions in the requirements on both these levels.

Second (2), we propose a specification method where we combine behavior-driven development (BDD) and test-driven development (TDD) with the scenario-based modeling technique. This enables the iterative specification of system features and usage scenarios to document stakeholder expectations and generate tests steps, which subsequently drive the scenario-based modeling of the system specification.

While numerous approaches exist that suggest using formal scenario models to bridge the gap from informal requirements to the implementation of software-intensive systems [5,13,28,29], the particular contribution of this paper is the extension of scenario-based modeling and programming techniques based on LSC Play-Out [13] and behavioral programming (BP) [15] with BDD and TDD. Enabling this combination of agile development techniques with scenario-based requirements modeling addresses the *coverage and sampling concerns* in scenario-based requirements engineering [28]: by connecting features with tests (BDD), and tests with the scenario-based requirements model (TDD), we can ensure that every feature is modeled by an appropriate set of scenarios, and that these scenarios are validated by an appropriate set of tests.

We asses the applicability with a proof-of-concept e-mobility application and provide a demonstration tool[1, 2] to enable others to use, evolve, and evaluate our approach.

Structure: We describe background in Sect. 2, the scenario-based requirements specification method in Sect. 3, and the proof-of-concept application in Sect. 4. We report related work in Sect. 5 and conclude in Sect. 6.

2   Background

2.1   System of Systems Engineering (SoSE)

For the description of System of Systems (SoS) no generally valid definition yet exists [1,26]. Hence, a distinction between complex monolithic systems and SoS is often made by the system characteristics. Therefore Maier describes five key characteristics of SoS [23]: (1) *Operational Independence*: Each system that is part of the SoS is independent and can perform a meaningful task, even if it is not integrated into the SoS. (2) *Managerial Independence*: The individual systems are self-administered and individually managed. Consequently they collaborate with the other systems of the SoS, but they operate independently. (3) *Geographic Distribution*: The individual systems of the SoS are distributed over large spatial distances, which means that the exchange of information between the individual systems is of primary importance for collaboration. (4) *Evolutionary Development*: The objectives and functionality of an SoS can change constantly, as they can be added, modified or removed based on experience. Therefore an SoS never appears to be fully completed. (5) *Emergent Behavior*: By the collaboration of the individual systems, a synergism is achieved in which the SoS fulfils a purpose that cannot be achieved by or attributed to any of the individual systems.

---

[1] https://bitbucket.org/crstnwchr/besos (includes the proof-of-concept example).

[2] https://bitbucket.org/jgreenyer/smlk/ (required to build the example project).

These characteristics have a strong influence on the SoS development. To support a structured SoS development Dahmann et al. [4] describe the differences between systems engineering (SE) and SoS engineering (SoSE). Accordingly, SE and SoSE both start with identifying and understanding user capability objectives in order to derive technical requirements for the system to be developed. In SE we subsequently continue with a top-down requirements decomposition and system design, with clear responsibilities in the management and engineering of the system [11]. In SoSE, by contrast, the identified objectives and requirements serve as a basis for the development of new systems *or* the integration of existing systems to build the SoS. Particularly the operational and managerial independence of individual systems is challenging: the existing systems may also fulfill other purposes that may conflict with the SoS objectives and those of its CS. Therefore it is important to understand how the individual systems behave and how this behavior contributes to the overall SoS behavior.

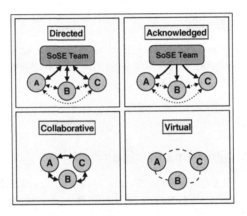

Fig. 1. Different types of SoS [4, 23]

When starting the SoS development it is important to categorize the SoS to be developed at an early stage because this has a significant influence on the RE approach that can be applied [4,25,26]. Figure 1 shows four different SoS types, initially introduced by Maier [23] and extended by Dahmann and Baldwin [4]: A *directed* SoS is designed for specific purposes. The individual systems have the ability to operate independently but are managed by a SoSE Team in a way that they fulfill a specific purpose. In an *acknowledged* SoS the SoSE Team recognises and defines a common purpose and goal, but the CSs retain independent control and goals. The continuous and evolutionary development of the common purpose is based on collaboration between the SoS and the CSs. In a *collaborative* SoS the individual systems are not bound to follow a central management, but voluntarily participate in a collaboration in order to achieve the SoS goal. A *virtual* SoS has neither a leading control nor a common goal. This leads to a high degree of emergent behavior where the exact means and structures that produce the functionality of the system are difficult to recognize and distinguish [26,27].

This paper focuses on acknowledged SoS and we introduce an example next.

2.2 Example of Application

To illustrate our approach, we introduce an e-mobility system of systems. In [21] Kirpes et al. introduce an architecture model that provides an integrative view on former separated areas of electricity, individual mobility, and information and communication technologies to realize future e-mobility SoS.

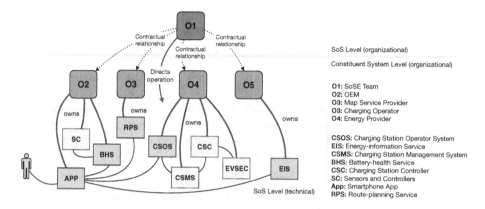

Fig. 2. Smart charging as an acknowledged SoS. Based on [25] and [21].

Based on the example defined in [21], Fig. 2 shows an SoS user who is interacting with an e-mobility SoS. The main interest of the user is to improve the e-mobility experience and to reduce its costs. These user interests are targeted by a high-level use case that describes how to create an optimized travel plan. At the beginning the user enters travel preferences like start and destination into the smartphone app (APP). The APP then requests further data from other systems that are necessary for the calculation of an optimized route. For example, GPS data of possible routes are requested from a route-planning service (RPS). Usage data of available charging points along the routes are provided by a charging station operation service (CSOS). Recommendations for a battery-saving charging process are provided by a battery-health service (BHS). And information on current electricity prices in the region is provided by the energy-information service (EIS).

These different systems, which are required to provide information to calculate an optimal route based on user preferences, are developed and managed by four different system owners (OEM, Map Service Provider, Charging Operator, and Energy Provider). We also see one SoSE Team which defines the overall SoS functionality, directs the operations, and has a contractual relationship with the owners of the CSs. According to [23] and [4] this example has the characteristics of an acknowledged SoS: we have recognised requirements, objectives and responsibilities on the SoS level and a contractual relationship between the SoSE Team and the individual constituent systems owner. However, the constituent systems keep their own management, funding and development approaches (cf. [25]).

2.3 Scenario Modeling Language for Kotlin (SMLK)

SMLK is a Kotlin-based implementation of the Behavioral Programming (BP) paradigm [15]. In BP, a program consists of a number of *behavioral threads*, which we also call *scenarios*. Scenarios are loosely coupled via shared events and can model individual behavioral aspects or functional requirements of a system. Scenarios can *request* events that shall happen, be *triggered by* or *wait for* events requested by other scenarios, or (temporarily) *forbid/block* events. During execution, the scenarios are interwoven to yield a coherent system behavior that satisfies the requirements of all scenarios.

Listing 1.1 shows two SMLK scenarios that can be represented graphically as shown in Fig. 3. Both scenarios are triggered by the event of a user entering the travel preferences in the app. This event is modeled as an interaction event of the object user sending the object app a message addTravel Preferences. In the first scenario, the parameters fromLoc and toLoc are variables bound to the parameter values carried by the triggering event when the scenario is triggered and initialized. The SMLK code in the listing shows this binding of the parameter values explicitly (lines 2 and 3). The second scenario is triggered by the same event, but does not use the parameter values; the sequence diagram expresses this by using asterisks.

After the trigger event, the first scenario requests that the app sends the Route Planning Service (rps) a message to calculate the route between fromLoc and toLoc , and then requests that the rps shall respond with a route. Then the app shall optimize the route and show it to the user.

The second scenario describes the interaction of the app and the Charging Station Operating System (csos) . After the triggering event, the scenario requests that the app sends the csos a request to send GPS position data of available charging stations. The scenario then requests that the csos shall respond with such a list. This interaction must happen before the app optimizes the route, i.e., the event app.optimizeRoute() is blocked until the second scenario terminates; only then can be first scenario proceed.

In these example scenarios, the route details and charging location list contents are not relevant, so mock instances are created by helper functions. When at a later point the behavior is refined, these parameter values may be replaced by other values, e.g., a detailed and correct route may be calculated elsewhere. The scenario method requestParamValuesMightVary allows us to request events with supplied default parameter values, but it will accept also events sent between the same objects, and with the same signature, but with different parameter values.

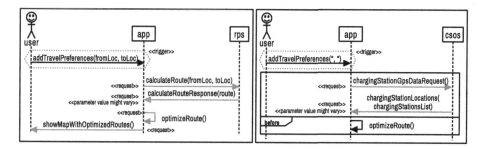

Fig. 3. Graphical representation of the SMLK scenario in Listing 1.1

```
1   scenario(user sends (app receives App::addTravelPreferences)){
2       val fromLoc = it.parameters[0] as String
3       val toLoc = it.parameters[1] as String
4       request(app sends rps.calculateRoute(fromLoc, toLoc))
5       val route = createMockRoute()
6       requestParamValuesMightVary(rps sends app.calculateRouteResponse(route)
7       request(app.optimizeRoute())
8       request(app sends user.showMapWithOptimizedRoute())
9   },
10  scenario(user sends (app receives App::addTravelPreferences)){
11      scenario {
12          request(app sends csos.chargingStationGpsDataRequest())
13          val chargingStationsList = createMockChargingStationsList()
14          requestParamValuesMightVary(csos sends app.considerChargingStationLocations(
                chargingStationsList))
15      }.before(app.optimizeRoute())
16  }
```

Listing 1.1. Example scenario from the e-mobility system specification

3   Scenario-Based Requirements Specification in a System of Systems Context

To develop an SoS, usually existing systems are integrated by new systems to comprise a new SoS. While the new systems may be under a direct managerial and operational control, existing systems may be under the managerial and operational control of another organization. Over time, systems that are under external control may change, which leads to the necessity to continuously (1) analyze how the changes in one system impact the SoS functionality, and (2) how other systems may have to be adapted to ensure that the SoS functionality can still be provided. This requires the SoSE team to continuously analyze, specify, and align requirements across different hierarchy levels.

Our scenario-based requirements specification approach supports an iterative and integrated behavior modeling and analysis on the SoS and CS level.

Based on the definitions in [14] we introduce the term *inter-system scenarios* to model the behavior on the SoS level and *intra-system scenarios* to model the CS behavior. Also we show how both views can be integrated to allow for the joint execution and testing of the integrated SoS and CS behavior.

3.1   Inter-system Scenarios

The goal of modeling inter-system scenarios is to conceive an validate how SoS use cases can be realized by the interaction of users, existing systems, and new systems to be developed. The inter-system scenario modeling process starts by defining the use cases, the structural SoS architecture, and then detailing and validating the use cases using scenarios and repeated simulation.

When modeling this behavior, certain assumptions are made about the behavior of the existing systems, possibly based on available documentation or communication with experts from the respective organizations.

Two exemplary inter-system scenarios are already introduced in Listing 1.1, where we first modeled the interaction between the app, rps and the SoS user, and in the second scenario, between the user, app and csos. In this example, we see that we are able to model the interaction between selected systems, where new requirements can be considered by iteratively adding new scenarios to the *SoS scenario specification*. By adding these inter-system scenarios the introduced modeling concepts allow to focus on a high level system interaction; Although we are able to partly ignore specification details (e.g. exact route information in Listing 1.1 line 14), we are able to execute and validate the interaction between the CS. This supports on the SoSE team to get a better understanding of the overall system behavior.

3.2   Intra-system Scenarios

Once a satisfactory concept of the inter-system behavior is established, the inter-system specification must be supplemented and refined in two ways: First (1), it is necessary to specify the behavior of the existing systems in more detail in order to validate whether the inter-system interaction behavior is indeed aligned with the behavior of the existing systems. Second (2), the behavior of the new systems to be developed must be detailed, possibly detailing their component structure and internal interactions, in order to provide a thorough basis for their development.

Our approach supports modeling the behavior on this more detailed hierarchy level with scenarios as well, and even to integrate their execution in order to simulate and validate behavioral requirements consistency across the different hierarchy levels.

To better distinguish between these two hierarchy levels, we distinguish the *inter-system* level and *intra-system* level as outlined in Fig. 4. The SoS scenario specification is located in the inter-system view, and individual CS scenario specifications are located in the intra-system view. When defining the internal behavior of a selected CS, we switch the perspective from the SoSE Team to a systems owner who is responsible for the development of a system. This can be e.g. the map service provider who is responsible for the development of the rps (see Fig. 2).

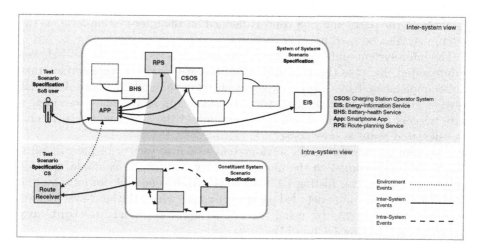

Fig. 4. Inter- and intra-system view to continuously concretise requirements on CS level, while also considering the overall SoS behavior.

The intra-system scenarios are added to an individual CS scenario specification, with the goal to model requirements which are needed to build the CS and its subsystems. One example intra-system scenario is shown in Listing 1.2.

```
1   scenario(routeRequester sends(rps receives Rps::calculateRoute)){
2       val fromLocString = it.parameters[0] as String
3       val toLocString = it.parameters[1] as String
4       request(rpsController sends gpsService.getLocations(fromLocString, toLocString))
5       val fromLoc = getLocation(fromLocString)
6       val toLoc = getLocation(toLocString)
7       request(gpsService sends rpsController.locations(fromLoc, toLoc))
8       request(rpsController sends routePlaner.calculateRoute(fromLoc, toLoc))
9       val route = calculateRoute(fromLoc, toLoc)
10      request(routePlaner sends rpsController.calculatedRoute(route))
11      request(rpsController sends routeRequester.calculateRouteResponse(route))
12  }
```

Listing 1.2. CS scenario specification of the RPS

The scenario specifies how the internal components of the rps (rpsController, gspService, and routePlanner) interact when receiving a request to calculate a route. Eventually (line 11), the calculated route will be returned to the requesting object.

When looking at the scenario in more detail, we see that the scenario is triggered when a `routeRequester` sends the `rps` the message `calculate Route`. This event is requested on the inter-system level, see the first scenario in Listing 1.1 (line 4).

One difference is, however, that in the intra-system scenario, we abstract from the app as being the source of the `calculateRoute` request (and the recipient of the route as a reposonse, see line 11). Instead, we assume that there is an abstract external route-requesting entity that requests a route to be calculated by the rps. We do this to separate the intra-system specification of a system from the particular SoS context defined on the inter-system level, as the system may also be used in other contexts.

The inter-system and intra-system level scenario execution can nevertheless be integrated, because the type of `routeRequester` is an interface that is also implemented by `app` (without showing the code in more detail for brevity). Hence it is possible that the event of the app requesting to calculate a route triggers the scenario shown here, and indeed the app would then receive the calculated route as a response.

The event parameters on the intra-system level may vary or be more detailed than the values assumed on the inter-system level where, for example, we used simple mock values (see Listing 1.1, lines 5 and 13). It is possible for intra-system scenarios to provide more detailed parameter values where the inter-system level scenarios request events by using the `requestParamValuesMightVary` command. (see Listing 1.1 line 14).

3.3   Specification Method

To support the requirements engineer in modeling system requirements with SMLK, we propose an iterative method based on agile techniques. Figure 5 shows an overview of the single steps. We start with the specification of the inter-system behavior by applying the BDD approach. Here we first define the expected system behavior from the SoS user perspective. Therefor we create a *SoS feature specification* where each feature is defined by one or more *usage scenarios* written in the gherkin syntax[3]. Listing 1.3 shows a first feature specification that describes a user interaction with the app. On this hierarchy level, the SoS feature specification allows the SoSE team to define what is expected from the SoS and to document this expectations in a comprehensible form.

```
1   Feature: Retrieve travel preferences and display optimized route
2
3       Scenario: Add travel preferences to the app
4           When the SoS user adds travel preferences to the app
5           Then the app displays a set of optimized routes
```

Listing 1.3. Initial feature specification including a usage scenario to describe the user interaction with the SoS.

Based on this SoS feature specification we generate test skeletons as shown in Listing 1.4. These test skeletons are then used to drive the modeling of the inter-system behavior. To support a structured and iterative modeling of system requirements, we embed the Test-Driven Scenario Specification (TDSS) [32] into the BDD approach. In this way, we combine the comprehensible specification of expected system behavior with the formal and scenario-based modeling of system requirements.

---

[3] https://cucumber.io/docs/gherkin/.

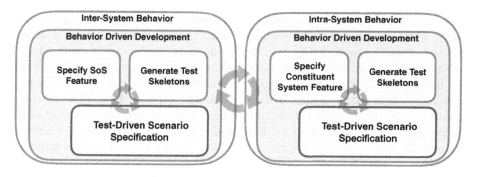

Fig. 5. Continuous and iterative scenario specification

```
1  When("^the EV user adds travel preferences to the App$") {
2      //implement here
3  }
4  Then("^the App displays a set of optimized routes$") {
5      //implement here
6  }
```

Listing 1.4. Generated test steps.

The TDSS approach includes the steps outlined in Fig. 6. In the first step we extend the generated test skeletons (1). Here, we e.g. model that the user adds travel preferences to the app (Listing 1.5 line 2) and eventually receives a map with optimized routes (line 5). After we added these functions we execute the SoS feature specification (2) whereupon the single test steps and finally the events within the test steps are executed. At this point in time we did not model the inter-system behavior and consequently the test fails, because the app will not send the optimized route to the SoS user as expected in line 5.

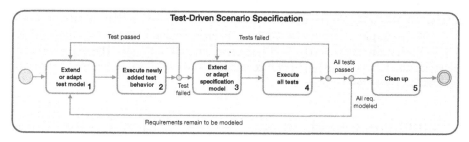

Fig. 6. Test-driven scenario specification (TDSS) [32]

```
1  When("^the EV user adds travel preferences to the App$") {
2      trigger(user sends app.addTravelPreferences("Dortmund", "Paderborn"))
3  }
4  Then("^the App displays a set of optimized routes$") {
5      eventually(app sends user.mapWithOptimizedRoutes())
6  }
```

Listing 1.5. Generated test steps.

Therefore we extend our SoS scenario specification with the inter-system scenarios (3) which we already introduced in Listing 1.1. We then run the test again to ensure that the modelled system requirements meet the expectations (4). If we have modeled additional tests in previous iterations, we now run them as well to ensure that there are no unexpected interactions between the individual tests and system requirements. If there are more requirements that need to be modeled, we perform further iterations. When all requirements on the SoS level known at this time have been modeled and tested, the SoS feature specification can be cleaned up. Afterwards the detailed specification of selected systems under development follows.

This iterative approach supports the modeling of the interaction of all CSs within the SoS. In this way we are able to iteratively document the expectations from an SoS user perspective and model and test the interaction between the CSs. Thereby new systems and behavior can be added as needed to realize the expected behavior. When we have gone through several iterations, the SoSE team gets a better understanding of which systems are needed and what information these systems have to exchange with each other. Subsequently we can switch to the intra-system level and focus on the requirements specification for a selected CS within the SoS. Based on our example outlined in Fig. 2 we now switch from the SoSE team perspective to e.g. the perspective of the map service provider, who is responsible for the development of the rps. As shown in Fig. 5 we execute the same specification method, but we create an independent *CS feature specification*, generate independent test steps and create an CS scenario specification. This allows the independent specification and modeling of the requirements for the CS, which addresses the managerial, operational and evolutionary independence of systems in an SoS. In this way, system requirements can be specified without seeing the system in an SoS context. But, at the same time, both views can be integrated (as described in Sect. 3.2), which allows the joint execution of the SoS behavior and the internal behavior of single already specified systems. In this way it's possible to detect contradictions between requirements on both levels. For example, if requirements have been specified at CS level that appear to have nothing to do with the SoS behavior but still influence the expected SoS behavior, the joint execution of the scenario specifications can be used to detect and resolve these dependencies.

4   Proof of Concept

To assess the applicability of our approach we integrated SMLK with the BDD tool Cucumber and executed the previously described specification method based on the example introduced in Sect. 2.2.

On SoS level we started with the feature specification as already shown in Listing 1.3. Subsequently we generated the test skeletons and added the SMLK events as shown in Listing 1.5. Following the TDSS approach we executed the SoS feature specification (Step 1 in Fig. 6) and got a failed test result as shown in Fig. 7. Subsequently we extended the SoS scenario specification as shown

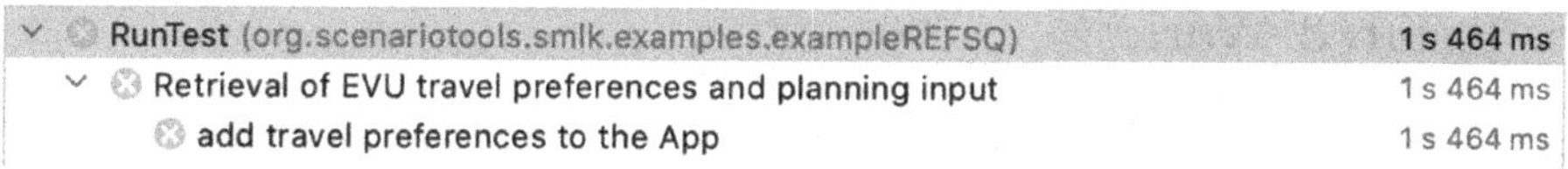

Fig. 7. First TDSS run on SoS level

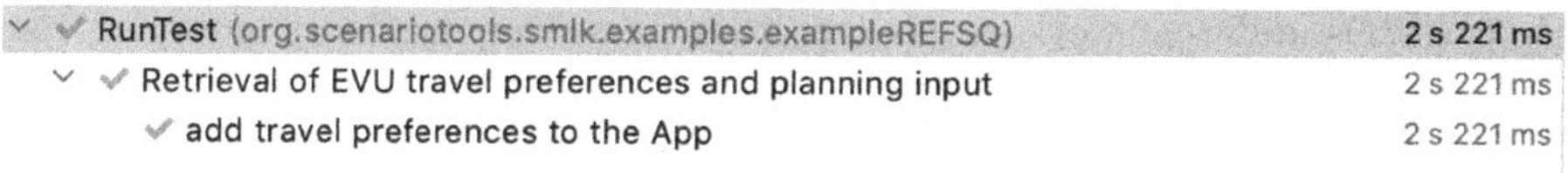

Fig. 8. Execute tests after adapting the SoS scenario specification

in Listing 1.1 to specify the SoS behavior. After we added these scenarios we executed the test again and finally received the expected event, resulting in a positive test result as shown in Fig. 8.

After we successfully defined a first interaction on inter-system level, we switched to the intra-system level and added a CS scenario specification to model the internal behavior of the rps as shown in Listing 1.2.

Now we executed the same SoS feature again resulting in a negative result, because the rps internal behavior was not yet specified and hence the CS scenario program didn't send the `calculateRouteResponse(route)` message to the app.

To fix this we executed the TDSS process within the intra-system view, based on the CS feature specification shown in Listing 1.6.

```
1   Feature: Calculate route - RPS
2
3       @RpsSystem
4       Scenario: Calculate route based on user travel preferences
5       When the app sends travel preferences to the rps
6       Then the rps responds route information including gps data
```

Listing 1.6. Feature on CS level

Finally we got passed test results again, but now we also considered the rps internal behavior specification. And, by using *tags* within the different feature specifications (e.g. `@RpsSystem`) and by applying the concepts described in Sect. 3.2, we were not only able to validate the integrated SoS and CS behavior, but we also could independently test the requirements of single CS.

To allow others to use, validate and evolve our approach, we describe the architecture and functional principles of the developed tool in [31] as a companion to this paper. Here, we also describe the method we outline in Fig. 5 in more detail. And we provide information about the necessary resources[4,5,6,7] to build and execute the example we use in this paper.

---

[4] https://bitbucket.org/crstnwchr/besos/.
[5] https://bitbucket.org/jgreenyer/smlk/.
[6] https://cucumber.io.
[7] https://www.jetbrains.com/idea/.

5   Related Work

In this paper we use SMLK, which was extended to support an iterative and continuous modeling of system behavior in an SoS context. This modeling language is based on Live Sequence Charts (LSCs) [6]. A recent LSC variant are Modal Sequence Diagrams (MSDs) [12]. By modeling behavioral requirements with the help of MSDs, different works argue that this formal requirements modeling can increase the requirements quality (e.g. [10,19]), but these approaches are based on traditional SE and do not consider the SoS characteristics and their impact on the requirements specification.

Harel et al. describe an extension to behavioral programming that allows the integration of behavioral programs that operate on different hierarchy levels and time scales [16]. Indeed, we also use this approach to integrate different SMLK scenario programs that execute the behavior on the inter- and intra-system level.

Simulation-based analysis and design is commonplace in cyber-physical systems of systems, e.g. using actor-oriented frameworks or co-simulation [9,22]. We aim to provide similar means for the thorough specification and analysis of *requirements* of systems of systems. To the best of our knowledge, this is a new approach.

Other works address model-based RE in the SoS context. Holt et al. describes an ontology for model-based SoS requirements engineering [18]. Albers et al. show how SoS requirements can be specified based on use-cases and sequence diagrams within SysML [2]. However, an early, iterative and formal specification of requirements, with the goal to execute and test these requirements specifications is not considered in these approaches.

6   Summary and Outlook

In this paper, we propose a technology to continuously model behavior requirements in an SoS context. Our approach supports requirements engineers in the iterative specification, modeling and testing of requirements. With the use of SMLK, the system behavior can be modeled textually through scenarios. This scenario-based modeling is close to how engineers communicate system behavior and hence enables a feasible formalization of requirements. To further support and structure the formalization process, we integrated SMLK with agile techniques and appropriate tooling. This fosters the iterative formalization, and by testing the formalized requirements specifications, we get early feedback about the expected system behavior and possible contradictions in requirements. Due to the proposed coupling of inter- and intra- system scenarios, we are also able to execute and test the system behavior on different hierarchy levels. And by integrating the BDD tool cucumber, we are able to specify the expected system behavior with the help of features and usage scenarios written in natural language, which supports the communication of expected system behavior in a multi-disciplinary development team.

For future work, we plan to integrate our previous work [32] and the modeling concepts shown in this paper with an automated test case creation proposed in [8] to further reduce the modeling effort. Also, as already started in previous work [33], we plan to integrate the results of this paper in an automotive development process and validate the applicability within an ongoing research project. As shown in [32], we are able to find contradictions in automotive requirements specifications, but the open questions are if the approach is scalable and whether the effort for the requirements modeling is justified.

Another possible direction for future work is focusing on stakeholder needs in a SoS context. In this paper we already integrated the BDD approach to validate requirements and align stakeholder expectations. This could be done more systematically by integrating goal modeling approaches [3].

References

1. Albers, A., Mandel, C., Yan, S., Behrendt, M.: System of systems approach for the description and characterization of validation environments. In: Proceedings of International Design Conference, DESIGN, vol. 6, pp. 2799–2810 (2018). https://doi.org/10.21278/idc.2018.0460
2. Albers, A., Kurrle, A., Moeser, G.: Modellbasiertes Anforderungsmanagement von Systems-of-Systems am Beispiel des vernetzten Fahrzeugs. In: Tag des Systems Engineering (TdSE), Bremen, 4–12 November 2014. Hrsg.: M. Maurer, pp. 373–382. Hanser, München (2015)
3. Aydemir, F.B., Dalpiaz, F., Brinkkemper, S., Giorgini, P., Mylopoulos, J.: The next release problem revisited: a new avenue for goal models (2018). https://doi.org/10.1109/RE.2018.00-56
4. Dahmann, J.S., Baldwin, K.J.: Understanding the current state of US defense systems of systems and the implications for systems engineering. In: 2008 2nd Annual IEEE Systems Conference, pp. 1–7 (2008)
5. Damas, C., Lambeau, B., van Lamsweerde, A.: Scenarios, goals, and state machines: a win-win partnership for model synthesis. In: Proceedings of the 14th ACM SIGSOFT International Symposium on Foundations of Software Engineering, SIGSOFT 2006/FSE-14, pp. 197–207. Association for Computing Machinery, New York (2006). https://doi.org/10.1145/1181775.1181800
6. Damm, W., Harel, D.: LSCs: breathing life into message sequence charts. Formal Methods Syst. Des. **19**, 45–80 (2001). https://doi.org/10.1023/A:1011227529550
7. Fernández, D.M., Wagner, S.: Naming the pain in requirements engineering. Empirical Softw. Eng. 183 (2013). https://doi.org/10.1145/2460999.2461027
8. Fischbach, J., Vogelsang, A., Spies, D., Wehrle, A., Junker, M., Freudenstein, D.: SPECMATE: automated creation of test cases from acceptance criteria. In: Proceedings - 2020 IEEE 13th International Conference on Software Testing, Verification and Validation, ICST 2020, pp. 321–331 (2020). https://doi.org/10.1109/ICST46399.2020.00040
9. Fitzgerald, J., Pierce, K., Larsen, P.G.: Co-modelling and co-simulation in the engineering of systems of cyber-physical systems. In: 2014 9th International Conference on System of Systems Engineering (SOSE), pp. 67–72 (2014). https://doi.org/10.1109/SYSOSE.2014.6892465
10. Fockel, M., Holtmann, J., Koch, T., Schmelter, D.: Formal, model- and scenario-based requirement patterns. In: 6th International Conference on Model-Driven Engineering and Software Development (2016). https://doi.org/10.5220/0006554103110318

11. Gausemeier, J., Moehringer, S.: VDI 2206- a new guideline for the design of mechatronic systems. In: IFAC Proceedings Volumes, pp. 785–790. Elsevier (2002). https://doi.org/10.1016/s1474-6670(17)34035-1

12. Harel, D., Maoz, S.: Assert and negate revisited: modal semantics for UML sequence diagrams. In: Proceedings of the 2006 International Workshop on Scenarios and State Machines: Models, Algorithms, and Tools, SCESM 2006, pp. 13–20. ACM, New York (2006). https://doi.org/10.1145/1138953.1138958. http://doi.acm.org/10.1145/1138953.1138958

13. Harel, D., Marelly, R.: Specifying and executing behavioral requirements: the play-in/play-out approach. SoSyM **2**, 82–107 (2003)

14. Harel, D., Marelly, R., Marron, A., Szekely, S.: Integrating inter-object scenarios with intra-object statecharts for developing reactive systems. IEEE Des. Test 1–19 (2020). https://doi.org/10.1109/MDAT.2020.3006805

15. Harel, D., Marron, A., Weiss, G.: Behavioral programming. Comm. ACM **55**(7), 90–100 (2012). https://doi.org/10.1145/2209249.2209270

16. Harel, D., Marron, A., Wiener, G., Weiss, G.: Behavioral programming, decentralized control, and multiple time scales. In: Proceedings of the Compilation of the Co-Located Workshops on DSM 2011, TMC 2011, AGERE! 2011, AOOPES 2011, NEAT 2011, & VMIL 2011, SPLASH 2011 Workshops, pp. 171–182. Association for Computing Machinery, New York (2011). https://doi.org/10.1145/2095050.2095079

17. Hoehne, O.M., Rushton, G.: A System of Systems Approach to Automotive Challenges. SAE Technical Paper. SAE International (2018). https://doi.org/10.4271/2018-01-0752

18. Holt, J., Perry, S., Brownsword, M., Cancila, D., Hallerstede, S., Hansen, F.O.: Model-based requirements engineering for system of systems. In: Proceedings - 2012 7th International Conference on System of Systems Engineering, SoSE 2012, pp. 561–566 (2012). https://doi.org/10.1109/SYSoSE.2012.6384145

19. Holtmann, J., Bernijazov, R., Meyer, M., Schmelter, D., Tschirner, C.: Integrated and iterative systems engineering and software requirements engineering for technical systems. J. Softw.: Evol. Process **28**(9), 722–743 (2016). https://doi.org/10.1002/smr.1780. https://onlinelibrary.wiley.com/doi/abs/10.1002/smr.1780

20. INCOSE: INCOSE Systems Engineering Handbook: A Guide for System Life Cycle Processes and Activities. John Wiley (2015)

21. Kirpes, B., Danner, P., Basmadjian, R., Meer, H., Becker, C.: E-mobility systems architecture: a model-based framework for managing complexity and interoperability. Energy Inform. **2**(1), 1–31 (2019). https://doi.org/10.1186/s42162-019-0072-4

22. Lee, K., Hong, J.H., Kim, T.: System of systems approach to formal modeling of CPS for simulation-based analysis. ETRI J. **37**, 175–185 (2015). https://doi.org/10.4218/etrij.15.0114.0863

23. Maier, M.W.: Architecting principles for systems-of-systems. In: INCOSE International Symposium, vol. 6, no. 1, pp. 565–573 (1996). https://doi.org/10.1002/j.2334-5837.1996.tb02054.x

24. Ncube, C.: On the engineering of systems of systems: key challenges for the requirements engineering community. In: 2011 Workshop on Requirements Engineering for Systems, Services and Systems-of-Systems, RESS 2011 - Workshop Co-located with the 19th IEEE International Requirements Engineering Conference, pp. 70–73. IEEE (2011). https://doi.org/10.1109/RESS.2011.6043923

25. Ncube, C., Lim, S.L.: On systems of systems engineering: a requirements engineering perspective and research agenda. In: Proceedings - 2018 IEEE 26th International Requirements Engineering Conference, RE 2018, pp. 112–123 (2018). https://doi.org/10.1109/RE.2018.00021

26. Nielsen, C., Larsen, P., Fitzgerald, J., Woodcock, J., Peleska, J.: Systems of systems engineering. ACM Comput. Surv. **48**, 1–41 (2015). https://doi.org/10.1145/2794381

27. Odusd, A., Sse, T.: Systems Engineering Guide for Systems of Systems. Technical Report August, Office of the Under Secretary of Defense (2008). https://doi.org/10.1109/EMR.2008.4778760

28. Sutcliffe, A.: Scenario-based requirements engineering. In: Proceedings of the IEEE International Conference on Requirements Engineering, pp. 320–329 (2003). https://doi.org/10.1109/ICRE.2003.1232776

29. Whittle, J., Schumann, J.: Generating statechart designs from scenarios. In: Proceedings of the 22nd International Conference on Software Engineering, ICSE 2000, pp. 314–323. Association for Computing Machinery, New York (2000). https://doi.org/10.1145/337180.337217

30. Wiecher, C.: A Feature-oriented approach: from usage scenarios to automated system of systems validation in the automotive domain. In: ACM/IEEE 23rd International Conference on Model Driven Engineering Languages and Systems (MODELS 2020 Companion), Virtual Event, Canada (2020). https://doi.org/10.1145/3417990.3419485

31. Wiecher, C., Greenyer, J.: Besos: a tool for behavior-driven and scenario-based requirements modeling for systems of systems, preprint (2021)

32. Wiecher, C., Greenyer, J., Korte, J.: Test-driven scenario specification of automotive software components. In: 2019 ACM/IEEE 22nd International Conference on Model Driven Engineering Languages and Systems Companion (MODELS-C), Munich, Germany, pp. 12–17 (2019). https://doi.org/10.1109/MODELS-C.2019.00009

33. Wiecher, C., Japs, S., Kaiser, L., Greenyer, J., Dumitrescu, R., Wolff, C.: Scenarios in the loop : integrated requirements analysis and automotive system validation. In: ACM/IEEE 23rd International Conference on Model Driven Engineering Languages and Systems (MODELS 2020 Companion) (2020). https://doi.org/10.1145/3417990.3421264

Specifying Requirements for Data Collection and Analysis in Data-Driven RE. A Research Preview

Maurizio Astegher[1], Paolo Busetta[1], Anna Perini[2(✉)] ⓘ, and Angelo Susi[2] ⓘ

[1] Delta Informatica SpA, Trento, Italy
{maurizio.astegher,paolo.busetta}@deltainformatica.eu
[2] Fondazione Bruno Kessler, Trento, Italy
{perini,susi}@fbk.eu

Abstract. [**Context and motivation**] According to Data-Driven Requirements Engineering (RE), explicit and implicit user feedback can be considered a relevant source of requirements, thus supporting requirements elicitation. [**Question/problem**] Less attention has been paid so far to the role of implicit feedback in RE tasks, such as requirements validation, and on how to specify what implicit feedback to collect and analyse. [**Principal idea/results**] We propose an approach that leverages on goal-oriented requirements modelling combined with Goal-Question-Metric. We explore the applicability of the approach on an industrial project in which a platform for online training has been adapted to realise a citizen information service that has been used by hundreds of people during the COVID-19 pandemic. [**Contributions**] Our contribution is twofold: (i) we present our approach towards a systematic definition of requirements for data collection and analysis, at support of software requirements validation and evolution; (ii) we discuss our ideas using concrete examples from an industrial case study and formulate a research question that will be addressed by conducting experiments as part of our research.

Keywords: Data-Driven Requirements Engineering · User feedback · Goal-Question-Metric · Goal-Oriented Requirements Analysis

1 Introduction

Data-driven Requirements Engineering (DDRE) provides methods and techniques at support of software developers and analysts willing to exploit user feedback for eliciting, prioritising, and managing requirements for their software products [9]. RE research has devoted huge attention to automating DDRE, but several challenges remain to be addressed in order to better integrate DDRE into a continuous software development process, as discussed, for instance, in [6, 7, 11]. Exploiting user feedback in other stages of the software requirements lifecycle, beyond requirements elicitation, and enacting traceability of feedback

ⓒ Springer Nature Switzerland AG 2021
F. Dalpiaz and P. Spoletini (Eds.): REFSQ 2021, LNCS 12685, pp. 182–188, 2021.
https://doi.org/10.1007/978-3-030-73128-1_13

to software design artefacts are discussed in [6] as two aspects among others that need more attention by the research community. In our research, we focus on implicit user feedback, that is data generated during a usage session, and collected via dedicated monitoring mechanism [10]. While explicit user feedback, e.g. user reviews, is collected on dedicated channels or social media and, usually, can be requested after and independently of a system design, the collection of implicit feedback cannot be designed after that the system has been deployed. Indeed, developers and analysts risk to struggle to interpret what is available, or even worse miss opportunities of collecting the right data that would help validating to what extent the system they built meets its stakeholder's goals.

Our research objective is manyfold. At first, we aim at understanding what usage data should be collected and analysed for the purpose of system requirements validation and evolution in a DDRE approach. We then aim at defining a method for systematically specifying requirements for such data collection and analysis, which we call also *requirements for implicit user feedback management*. These requirements should be specified at design-time, linking to stakeholders' goals, and guide requirements validation and evolution once the software application has been deployed on an instrumented platform and accessed by its users, thus generating usage logs.

Motivations for this work derive from an industrial project in which a platform for online training was adapted as a citizen information service during the COVID-19 pandemic. Towards achieving our research objective, we first analyse examples from this project's case study. Moreover, we take inspiration from goal-oriented approaches for Business Intelligence, e.g. [2,8], and investigate whether concepts from Goal-Question-Metric (GQM) [3] can be exploited. We derive then a research question about the effectiveness of the proposed method that we aim at assessing with dedicated experiments in future work.

The rest of this short paper is organised as follows. We shortly recall background concepts of GQM and discuss related work in Sect. 2. We present the case study in Sect. 3, together with a discussion of examples of missing data collection and analysis. To tackle our research questions, we elaborate on our proposed approach in Sect. 4. Then, Sect. 5 concludes the paper highlighting ongoing and future steps in our research.

2   Background and Related Work

Goal-Question-Metric (GQM) is a top-down method for deriving and selecting a set of metrics to assess the achievement of high level goals [3]. A high level goal is decomposed into sub-goals. Questions referring to what could help stating that those goals are achieved are then identified. Metrics are derived, which individually or in an aggregated form can help answering each question. A simple example is depicted in Fig. 2 (left side). GQM has been introduced first in software engineering but it has been widely applied in different contexts, including business strategies assessment [4]. Our research applies GQM to functional and quality goals that represents users' requirements for a software application, with the aim of defining, in a top down breakdown, what data to collect and how to analyse them in order to get evidences about user's requirements satisfaction.

Implicit user feedback, as defined in [9,10], is related to business process analysis and process mining. While the objective of process mining is to extract knowledge about processes from transaction logs, aiming at detecting or preventing misbehaviour and monitoring process quality, in DDRE the analysis of implicit user feedback aims at eliciting new requirements and, in our research, at supporting requirements validation and evolution. In process mining, three main data analysis perspectives have been proposed, namely the process perspective, the organisation perspective and the case perspective [12]. In [1], a literature based meta-model is presented, which captures the most relevant knowledge elements that have been considered in process mining literature so far, including the notion of actor's goal. By contrast, this work considers the user's goals perspective as part of the software system requirements specification.

3   Case Study

During the state of emergency for the first wave of the COVID-19 pandemic, the local government of the Provincia Autonoma di Trento, Italy (PAT for short) needed a way to regularly update citizens (more than half a million inhabitants) about the prudent and legally permitted behaviours to follow and those that did not comply with health advice and norms. Browsing through cryptic and lengthy regulations looking for clues concerning a specific topic can be a tedious and difficult process, and people often preferred to interact directly with a PAT operator via telephone even for the simplest questions.

To lighten the workload of call centre operators, a Web system was created to present information, organised in guidelines, in an immediate and interactive way. For each guideline, a detailed description of the allowed behaviours was produced; each of the latter was then associated to a reference category (e.g. *Sports and outdoor activities*) and to additional keywords. Web users could search by category or by keywords, similar to what happens in search engines. In many cases, text-based guidelines were accompanied by infographics and exercises carried out with the ELEVATE platform [5]. Here, we focus on the latter.

ELEVATE is primarily used for creating interactive, multi-medial exercises to be embedded into online training courses. However, an exercise can be used also as a *communication tool* that allows its user to explore answers to questions. The COVID-19 exercises ask simple questions such as *Do you have to use public transport to reach your destination?* and show a set of predefined answers to choose from. The exercises then unfold by proposing alternative scenarios depending on which choices are taken, ending when these lead to situations of either correct or discouraged (if not prohibited) behaviours. The user is free to repeat any exercise and explore different scenarios. Intuitively, a ELEVATE exercise allowed to organise information so that only the relevant sections of the COVID-19 directives were progressively offered to the user, leading to simplicity and greater engagement when compared to the sequential reading of norms written in legal language. Indeed, positive feedback on the exercises was informally collected both from users and from the press office of the local government. Months after their release, we tried to check if this positive feedback was confirmed by the data made available directly from ELEVATE.

The platform collects data on the user behaviour, such as the speed of execution or the ability to follow the correct paths of an exercise, thus achieving predefined educational objectives. In training, this data is used to evaluate the performance of individual and groups of students, and allows to tune the exercises e.g. by making situations harder or easier to interpret and offering alternatives to follow. Unfortunately, while these data gives us a very precise picture of the interests of the COVID-19 users (for instance, how many took a certain path within an exercise, i.e. how many were interested to specific situations), they do not allow to answer a few basic questions: (i) Did users find the information they were looking for? (ii) Overall, was using the exercise a positive or a negative experience? (iii) Are there extensions or improvements of the system we should consider after looking at the users' behaviours?

4   User Feedback Requirements and Related Tasks

An overview of the role of the requirements specification and validation tasks we propose to explicit when integrating DDRE into software development is sketched in Fig. 1. We distinguish between design-time, and post-deployment (run-time) tasks. At design-time, once key stakeholders have been identified together with their main goals, in parallel to the specification of functional and quality requirements of the intended software application or service, *GQM analysis* should be performed with the aim to define metrics, and corresponding range of values, which will help assess to what extent the running application achieve the stakeholders' goals. At run-time, i.e. once the software application has been deployed on a platform instrumented with data collection and analysis mechanisms, and it is accessed by its intended users, two tasks can be performed, namely *Requirements validation* and *Requirements evolution*. Concerning Requirements validation, the implemented mechanisms for data collection

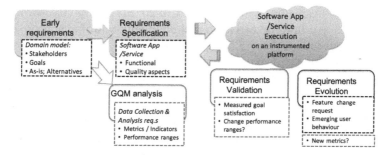

Fig. 1. The envisioned process with new tasks and artifacts highlighted in red. (Color figure online)

and analysis can help requirements engineers to evaluate if (and to what extent) the software application meets the stakeholders goals, and in addition to validate knowledge used to define the data collection and analysis requirements, as for instance value ranges of the indicators used in the metrics for goal assessment. As for Requirements evolution, apart from feature change requests that can be collected through explicit user feedback, and ideas for new requirements that can be suggested via analysing session logs, we foresee the possibility to elicit from implicit user feedback also ideas for new metrics, e.g. by aggregating indicators, by leveraging on process mining techniques.

In the rest of this section, we present some illustrative examples taken from our case study on how GQM analysis can help specifying user-feedback requirements, which can guide requirements validation.

Specification of Requirements for Data Collection and Analysis. As an example of use of GQM for specifying requirements for data collection and analysis we consider a key stakeholder in our case study, namely PAT, and its main goal *G1: Communicate Citizens new rules, with ELEVATE exercises.*

An example of GQM for goal G1 is summarised in Fig. 2, left side. The questions *Q1: Did Citizen access the ELEVATE exercises?* and *Q2: Did Citizen completed successfully the exercises?* have been associated to the goal G1. Corresponding identified metrics are *M1: Number of Citizens who accesses to an exercise in a time-period, M2: Average time spent by Citizens on an exercise in a time-period, M3: Percentage of Citizens who completed successfully an exercise in a time-period.* Figure 2, right side gives an example of how these metrics can be modelled using the approach proposed in [2], which would be particularly convenient when requirements analysis is performed using goal models. The example focus on the metric M3. A value of M3 above a given threshold (e.g. 70%) is an evidence for G1 satisfaction. Note that the threshold value is part of these requirements definition and can depend on the specific application domain.

Requirements Validation. For the purposes of this paper, we focus on three ELEVATE exercises that were created at the end of the national lockdown period and advertised by means of press releases to local media. Topics and key concepts of the exercises were extrapolated from an ordinance, issued by the local government, which came into force on May 4, 2020: *Sports and outdoor activities* (in short, sports can only be carried out individually and, for non-professional athletes, only outdoors); *Use of protection equipment* (mandatory use of the mask throughout the region both outdoors and in closed places accessible to the public); *Travelling* (travel within the region is only allowed for reasons of health, work, necessity or visits to relatives). Usage data were collected up to May 18, the day on which a further ordinance entered into force which led to a substantial easing of the restrictions in place up to that moment. Table 1 shows (i) the number of citizens who have accessed each exercise; (ii) the number of citizens who have completed it; (iii) the number of citizens who have completed it with a positive outcome; (iv) the percentage of successfully completed exercises.

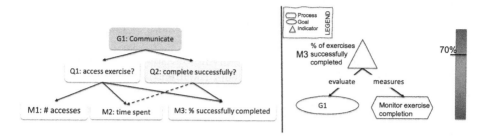

Fig. 2. Requirements expressed as GQM [3] and GO models [2].

Table 1. Values of indicators considered for the G1 validation example

Exercise name	Time window	Citizens	Completed	Compl. with success	Success%
Sports and outdoor activities	07/05–18/05	608	412	340	83%
Use of protection equipment	07/05–18/05	243	151	118	78%
Travelling (first version)	07/05–13/05	2139	1291	1228	95%
Travelling (second version)	13/05–15/05	254	159	145	91%
Travelling (third version)	15/05–18/05	356	249	211	85%

According to the chosen metric (indicator), the goal of Communicating citizens new rules (G1) of our stakeholder PAT seems to be well satisfied. Of course other goals, such as Citizen's goal of acquiring information on new, updated rules in an easy and effective ways would need the definition of appropriate metrics to be operationalised into corresponding data collection mechanism. Finally, from the point of view of the developers, such collected data would need to be automatically analysed and displayed in a corresponding suitable dashboard.

5   Conclusion

In this research preview, we have introduced our research on exploiting implicit user feedback in DDRE for requirements validation and evolution. We have presented our twofold research objective, which concerns the identification of tasks and artefacts in a DDRE process where implicit user feedback is exploited, and the definition of a method for the specification of requirements for implicit user feedback management based on GQM.

Guided by this research goal, we envisioned a design-time task dedicated to the specification of requirements for data collection and analysis, and propose a method that builds on GQM. At run-time, a task devoted to requirements validation is proposed, as well as the analysis of data that could support identifying new GQM. We illustrated our ideas on a case study concerning a citizen information service for the COVID-19 pandemic regulations in an Italian region.

As future work, we will assess the proposed specification method considering the following research question: *RQ: Is the proposed method (based on GQM combined with goal-oriented requirements models) an effective method for specifying requirements for implicit user feedback management, at support of software validation, and requirements evolution?*

We plan to perform an experiment involving developers who will be asked to specify implicit user feedback management requirements with our approach. Among the metrics for this experiment we will consider perceived usefulness, as well as objective metrics, such as the number of requirements specified in a given time frame. The experiment will be conducted on the ELEVATE project first, and possibly on other project for generalisation purposes.

Acknowledgements. This work is part of the ELEVATE research project, which is funded by Provincia Autonoma di Trento, L.P. 6/1999.

References

1. Adamo, G., Di Francescomarino, C., Ghidini, C.: Digging into business process meta-models: a first ontological analysis. In: Dustdar, S., Yu, E., Salinesi, C., Rieu, D., Pant, V. (eds.) CAiSE 2020. LNCS, vol. 12127, pp. 384–400. Springer, Cham (2020). https://doi.org/10.1007/978-3-030-49435-3_24
2. Barone, D., Jiang, L., Amyot, D., Mylopoulos, J.: Reasoning with key performance indicators. In: Johannesson, P., Krogstie, J., Opdahl, A.L. (eds.) PoEM 2011. LNBIP, vol. 92, pp. 82–96. Springer, Heidelberg (2011). https://doi.org/10.1007/978-3-642-24849-8_7
3. Basili, V.R.: Goal question metric paradigm. Encycl. Softw. Eng. **1**, 528–532 (1994)
4. Basili, V., et al.: Aligning Organizations Through Measurement. TFISSSE. Springer, Cham (2014). https://doi.org/10.1007/978-3-319-05047-8
5. Dellagiacoma, D., Busetta, P., Gabbasov, A., Perini, A., Susi, A.: Authoring interactive videos for e-learning: The ELEVATE tool suite. In: Vittorini, P., Di Mascio, T., Tarantino, L., Temperini, M., Gennari, R., De la Prieta, F. (eds.) MIS4TEL 2020. AISC, vol. 1241, pp. 127–136. Springer, Cham (2020). https://doi.org/10.1007/978-3-030-52538-5_14
6. Franch, X., et al.: Towards integrating data-driven requirements engineering into the software development process: a vision paper. In: Madhavji, N., Pasquale, L., Ferrari, A., Gnesi, S. (eds.) REFSQ 2020. LNCS, vol. 12045, pp. 135–142. Springer, Cham (2020). https://doi.org/10.1007/978-3-030-44429-7_10
7. Groen, E.C., et al.: The crowd in requirements engineering: the landscape and challenges. IEEE Softw. **34**(2), 44–52 (2017)
8. Horkoff, J., et al.: Strategic business modeling: representation and reasoning. Softw. Syst. Modeling **13**(3), 1015–1041 (2012). https://doi.org/10.1007/s10270-012-0290-8
9. Maalej, W., Nayebi, M., Johann, T., Ruhe, G.: Toward data-driven requirements engineering. IEEE Softw. **33**(1), 48–54 (2015)
10. Morales-Ramirez, I., Perini, A., Guizzardi, R.S.S.: An ontology of online user feedback in software engineering. Appl. Ontol. **10**(3–4), 297–330 (2015)
11. Perini, A.: Data-driven requirements engineering. the SUPERSEDE way. In: Lossio-Ventura, J.A., Muñante, D., Alatrista-Salas, H. (eds.) SIMBig 2018. CCIS, vol. 898, pp. 13–18. Springer, Cham (2019). https://doi.org/10.1007/978-3-030-11680-4_3
12. Van Der Aalst, W.M., et al.: Business process mining: an industrial application. Inf. Syst. **32**(5), 713–732 (2007)

Analysts' Competence and Training

SaPeer Approach for Training Requirements Analysts: An Application Tailored to a Low-resource Context

Jéssyka Vilela[1](✉) and Alessio Ferrari[2]

[1] Universidade Federal de Pernambuco (UFPE), Recife, Brazil
jffv@cin.ufpe.br
[2] CNR-ISTI, Pisa, Italy
alessio.ferrari@isti.cnr.it

Abstract. [**Context and Motivation**] Role-playing is a typical pedagogical strategy frequently applied in requirements engineering education and training (REET). The technique was proven to be successful for teaching different requirements engineering (RE) activities, and the SaPeer role-playing approach was recently proposed to train students in requirements elicitation interviews. SaPeer was shown to be effective and useful in the context of a *high-resource* RE module involving seven tutors, and a three-weeks individual assignment. [**Question/Problem**] RE lectures are frequently conducted as part of software engineering courses, or in short RE modules, and there is often limited time to teach RE in general, and interviews in particular. Therefore, SaPeer needs to be adapted to these constrained contexts, and adequately assessed. [**Principal idea/Results**] In this paper, we present the application of SaPeer to a *low-resource* context. We tailor the approach to a one-week group assignment, involving one tutor only, and we apply it to a class of 24 students. By comparing our results with the original study, we find that students struggle in similar areas, and especially in question omission and planning. A qualitative analysis of the feedback of the students shows the appreciation for the interview experience, and offers specific recommendations for improving the educational material. [**Contribution**] We contribute to the literature in REET with the first tailored application of SaPeer. Our study extends the scope of SaPeer and offers the possibility of adopting it in other constrained contexts.

Keywords: Requirements engineering · RE education and training · SaPeer Approach · Requirements Elicitation · Interviews

1 Introduction

In requirements engineering education and training (REET), it is common to use the *role-playing* technique for teaching students how to conduct requirements engineering (RE) activities [24,25,28,31,35]. With role-playing, students are normally asked to play the role of the requirements analyst in a fictional

© Springer Nature Switzerland AG 2021
F. Dalpiaz and P. Spoletini (Eds.): REFSQ 2021, LNCS 12685, pp. 191–207, 2021.
https://doi.org/10.1007/978-3-030-73128-1_14

project, so to have a hands-on experience of the difficulties of the profession, which requires a complex combination of rigorous planning, flexible execution, technical competence, communication abilities, and soft-skills. Role-playing has been demonstrated to be successful for training students in different RE phases, such as elicitation [4,10,12], analysis [1,23], and validation [3,21]. In the last years, requirements elicitation interviews received particular attention, given their prominence in the RE practice [7–9,11,14,19], and novel approaches for interview training were developed [5,12]. In particular, the SaPeer approach was proposed for training students in interviews with role-playing, self-assessment, and peer-review [12,13], and the authors shared the material to replicate and *adapt* the study [15]. Indeed, each educational context has its own peculiarities, characterized by the students' background, the educational goals prioritized by the teacher, and the availability of educational resources, such as number of tutors and amount of time. Therefore, staging an effective role-playing RE activity needs to account for the specific educational context. In particular, extensive training sessions in RE are often not possible, as RE modules are normally made of a limited amount of lectures, frequently included within larger software engineering courses [16,20].

In this paper, we propose an adaptation of the SaPeer approach. SaPeer was originally experimented in a high-resource RE course, with seven tutors playing the role of customer, and a three-weeks individual assignment. Here, we apply SaPeer to a low-resource context, which required tailoring the approach to the needs of the course. An application with some changes to adapt to different conditions may be classified as an *external differentiated replication* [2]. It is *external* since it was performed by an independent researcher, i.e., the first author. Besides, the second author, who is one of the SaPeer proponents, was only contacted after the execution of the study. It can be called *differentiated replication* since changes were performed in relation to the original study: design, hypothesis, and context. To avoid confusion, we use the term *application* instead of differentiated replication to refer to this study.

The main changes with respect to the original study are: the students work in groups instead of performing individual activities; the assignment lasts one-week; only one tutor (the professor) plays role of customer. The study involves 24 Brazilian students from a RE class in a Postgraduate Course about Software Testing.

We analyze the outcome of the work from a quantitative and qualitative point of view, and we compare the results with the original study. Our results show that both groups of students struggle in the areas of question omission and planning and tend to forget to ask about additional stakeholders and feature prioritization. Furthermore, time management and listening skills are particularly poor. On the other hand, the simple suggestion of asking a summary at the end of the interview—given as advice already before the first interview—allows students to correct this typical mistake beforehand. Our main contributions are as follows: (1) we provide one of the few cases of external application in REET and RE in general; (2) we show how SaPeer can be successfully adapted to a low-resource context, which is typical in REET [16,20]; (3) we confirm most of the results of the original study in terms of easiness and usefulness of SaPeer.

The paper is organized as follows. Related works are discussed in Sect. 2. A brief description of SaPeer, the rationale for tailoring it and how the tailoring was performed are described in Sect. 3. The research design is presented in Sect. 4. The results are described in Sect. 5 and Sect. 6 concludes the paper.

2   Related Work and Background

Requirements Engineering Education and Training (REET) is a lively area of research, with several relevant contributions addressing multiple phases of the RE process [1,3,4,10,12,21,23]. A systematic mapping study on the topic [26] identifies 79 high-quality research papers in the field already in 2015. The study highlights that requirements elicitation, which is the focus of our study, is addressed solely by 11% of the contributions and that the majority of the studies are either solution proposal or experience papers, including 34% of non-empirical studies. This suggests that the empirical maturity of the field is limited and calls for further contributions with a stronger experimental facet, especially considering *replications* or *tailored applications*, which are well-known pain points in RE research [6,22]. We are aware of a few replications in REET, such as: the work of Walia and Carver [33], with four experiments on requirements inspection; the one by Hadar et al. [18], concerned with the evaluation of requirements model comprehensibility; and the one of Spoletini *et al.* [30], about two experiments on interview review. Finally, the recent work of Rueda *et al.* [29] compares different requirements elicitation methods in a family of experiments involving the same instructor.

Role-Playing in REET. Many solution proposals in REET typically use *role-playing* as a pedagogical strategy for training students [17,27]. The seminal contributions by Zowghi and Parjani [35] are one of the first works that propose to apply role-playing to REET, and discusses lessons learned, covering aspect related to student engagement and corrective feedback. Svensson and Regnell [31] make a step forward, empirically showing the effectiveness of the role-playing pedagogical strategy. Similarly, Nkamaura and Tachikawa [24] confirm that requirements modeling skills can be improved through role-playing, while Vilela and Lopes [32] show improvements in requirements elicitation and communication abilities. Finally, the results of Ouhbi [25] show the adequacy of role-playing as a tool for teaching RE even with limited time resources, as in our case.

Mistakes of Student Analysts. The investigation of common mistakes of students acting as analysts in role-playing interviews is presented in previous works by the authors of the SaPeer approach [4,5,10]. Specifically, Donati et al. [10] identify nine main communication mistakes of 36 student requirements analysts in a case study. Bano *et al.* [4,5] present a list of 34 detailed mistakes that novices perform in requirements elicitation interviews classified into seven high-level categories (order of interview, planning, communication skills, etc.).

Based on that work, Ferrari *et al.* [12] introduce the SaPeer approach for training student analysts and correct typical mistakes. The work shows that the proposed method is effective in reducing students' mistakes and is considered useful.

Our paper replicates the study of Ferrari *et al.* [12], to investigate the applicability of the SaPeer approach in a low-resource context. With respect to other works in REET, we contribute with one *external* application in the field, led by an independent author that was not involved in the initial study, and with a focus on elicitation. As an application of an existing study, our work reinforces the empirical grounding of REET, extends the scope of applicability of SaPeer, and suggests useful improvements for the educational material provided.

3   Tailored SaPeer

In this section, we provide a brief description of SaPeer, the rationale for tailoring it and how the tailoring was performed.

3.1   The SaPeer Approach

The SaPeer pedagogical approach aims to foster *experiential learning* by letting students perform a role-playing interview with a fictional customer. Then, learning is further stimulated through reflection, by asking students to find mistakes in their own interview and in the interview of their peers [12]. The acquired ability is then practiced in a second interview.

The main steps of the approach are: (1) **Preliminary Training:** the students watch a video on how to conduct interviews; (2) **1st Interview:** the students act as analysts and a tutor plays the role of customer; (3) **Mistake-based Training:** the students watch a second video in which the common mistakes presented by Bano et al. [4] are explained, and examples of erroneous behavior are given; (4) **Self-assessment:** the students listen to their interview, and answer a questionnaire with 32 statements concerning the occurrence of mistakes; (5) **Peer-review:** the students assess another student's interview; (6) **2nd Interview:** a second interview is carried out for further practice; (7) **Self-reflection:** the students answer a feedback questionnaire about the usefulness and easiness of the SaPeer approach. The videos, slides and questionnaires of the SaPeer approach can be found at [15].

3.2   Rationale for Tailoring the SaPeer Approach

The context in which SaPeer would be applied was an *online* RE class with 24 students in a graduate course about software testing. The classes had a duration of 15 h (total) distributed in five days with three hours each.

Besides having a reduced time schedule, no additional tutor was available, and all classes and assignments would be taught and graded by only the professor. Hence, there was a need of performing the first adaptation, which was conducting the interviews in *teams* considering that time was a major issue in this class. Accordingly, the students were divided into six groups of four members. The original study also foresees the possibility of conducting the interviews in groups if the scale is an issue in applying the approach. Furthermore, it was an opportunity for students to handle communication flaws between the project team and the customer, avoiding typical problems arising from the presence of one team member only interacting with the customer [11].

The reduced available time was also the reason for choosing reducing the time of interviews from 15 min as adopted in the original study to the 10 min adopted in this application.

The professor selected three projects, instead of the original two, to reduce the possibility of cheating. The projects are listed in Table 1. Similarly to Bano et al. [12], the task was collaborative. The students were expected to plan for the interview as a group and assigning among them the different tasks such as preparing questions, asking questions, taking notes, audio recording interviews, preparing minutes of meeting. It is important to highlight that self-assessment and peer-review questionnaires were filled only after the first interview. Since the class had a duration of one week only, it was not possible to quantitatively evaluate the actual improvement of the students. We mitigated this issue by evaluating their qualitative feedback.

Also, while in the original study the two interviews were about different products, in this study, the second interview was a clarification interview, and, therefore, the questionnaire may not be adequate to identify mistakes in this phase, as also noticed by Bano *et al.* [5].

Finally, another adaptation was the type of artifacts produced by the students. In our application, it was necessary that students train several artifacts. Hence, they produced other artifacts beyond user stories requested by the original study.

3.3   The Tailored SaPeer Approach

This work presents an external tailored application [2] of the study of Ferrari *et al.* [12]. The differences between the original study and this application are presented in Table 1.

We adapted the suggested timeline to apply the approach to the course duration (five days). The timeline was the following:

[Day 1]: A class about Introduction to RE. Students watched the "Preliminary Training" video on interviews. Students planned the interview and sent the script they intend to follow during the interview. Students executed the 1st interview. Students sent the meeting notes immediately after the completion.

[Day 2]: A class about Requirements Analysis and Specification. Students watched the "Mistake-based Training" video. Students answered the self-evaluation questionnaire. Students listened to other team interview (team with a different project than theirs) and answered the peer-review questionnaire.

Table 1. Comparison between the original study and this application.

Setting	Original study [12]	This application
Study goal	Evaluate the learning effect of the proposed approach and to acquire feedback on its usefulness and easiness	Replicate the activities performed by the experimental group to acquire feedback on usefulness and easiness of the approach
Research Method	Quasi-experiment with two treatments: SaPeer and practice-only	Application without performing a controlled experiment
Country	United States	Brazil
Type of class	Physical	Online RE class
# Participants	16 in experimental group and 22 in the control group	24
Students' profile	Graduate students where around 50% had previous experience in RE	Graduate students where only 8% had experience in RE
Time available to perform the activities	Three weeks	One week
#interviews	2	2
Duration of interviews	15 min	10 min
Team involved in the role of customer for the role-playing activity	Seven tutors	One professor
Type of interview	Individual	Groups of four students
Language of the classes	English	Portuguese
Projects	Cool Ski Resorts: an information system to manage a chain of three Ski resorts [15], Nancy/Jim's Salon: an information system to manage a hair dressing shop [15]	Cool Ski Resorts, Nancy/Jim's Salon and Emergency medical response system: an application of smart items in the field of sensors network (designed by the instructor)
Artifacts produced by the students	User stories	Problem and business description, user stories, specification of requirements (functional with priority suggestion, non-functional and business rules), use case diagram, specification of 3 use cases, and one test case for a use case as long as it has at least one alternate flow

[Day 3]: A class about requirements types: functional, non-functional, and business rules. Production and submission of the requirements in the form of user stories.

[Day 4]: A Class about validation and requirements management. Students executed the 2nd interview on the same project. Production and submission of a list containing Functional requirements with priority suggestion, Non-Functional Requirements, and Business Rules.

[Day 5]: Students sent the requirements document containing the artifacts listed in Table 1 and they filled out the feedback questionnaire. Students presented the results of the elicitation to the professor and the class.

4   Research Design

In this work, we applied SaPeer to a low-resource context, and compared the obtained results with the original study. In the following, we outline research questions, data collection and data analysis procedures.

4.1   Research Questions

The following research questions motivated the conduction of this work:

RQ1: What are the most frequent mistakes performed by the students? We analyze the most common errors performed by the students considering the mistakes categories provided by the SaPeer approach and we compare the results of this study with the results of the original study, considering the average between self- and peer-reviews scores obtained for the first interview. This allows understanding whether mistakes are similar for students with different backgrounds and culture.

RQ2: What is the degree of easiness and usefulness of the approach from the viewpoint of the students? We collect students feedback regarding their opinion about easy of use and their perception of the utility of the SaPeer approach and we compare the results of both studies obtained from the feedback questionnaire.

RQ3: What are the benefits and challenges of SaPeer from the viewpoint of the students? To answer this question, we present a thematic analysis of the feedback questionnaire to qualitatively understand to which extent the students considered the approach effective.

4.2   Data Collection and Analysis Procedures

To collect the data to answer our research questions, we asked the students to fill the following questionnaires:

- **self-assessment (RQ1):** this questionnaire contains 32 statements, one for each mistake type described in the mistake-based training. For each statement, the student was required to provide a degree of agreement in a 5-point Likert Scale: Strongly Agree (5), Agree (4), Neutral (3), Disagree (2), Strongly Disagree (1).

- **peer-review (RQ1):** this questionnaire is similar to the self-assessment one, except for the formulation of the statements, which in this case are in third person.
- **feedback (RQ2, RQ3):** with this questionnaire, the students evaluate the usefulness and easiness of the approach, and provide comments on their experience. The students are asked to evaluate their usefulness on a 5-point Likert Scale: Extremely useful (5), Very useful (4), Moderately useful (3), Slightly useful (2), Not at all useful (1). Similarly for easiness: Very easy (5) to Very difficult (1). In addition, the students were required to comment on the effectiveness of their experience and recommend improvements.

To answer RQ1, we quantitatively analyze the answers to the self-assessment and peer-review questionnaires, and we compute their average for each mistake. This is compared with the average score obtained in the original study.

To answer RQ2, we analyze the answers to the feedback questionnaire about easiness/difficulty, and for each task, we compare the average scores with the original study. RQ3 was answered by performing a thematic analysis similar to the one presented in the paper of Ferrari *et al.* [13]. The themes were grouped into three categories: Challenges, Benefits, and Improvements.

4.3   Threats to Validity

As we performed an external application of the study of [4], we also inherit some of the threats to validity reported in their paper. Below, we analyze the threats to validity considering the classification of Wohlin *et al.* [34]. Considering *construct validity*, we mitigate threats related to the amount of mistakes by calculating the average between self-assessment and peer-review scores to reduce possible students' bias in assigning their scores. Since students performed group interviews, while they answered individual questionnaires, the actual self-assessment scores actually reflect a group score. By considering solely average values, we mitigate this issue.

To reduce threats to *internal validity*, a single tutor played the role of customer in all interviews. Hence, it was possible to provide similar answers to all groups. Besides ethical issues discussed below, a possible source of bias is the fact that the leader of the study, the course instructor, and the first author of the paper are the same person. However, we argue this threat is limited because SaPeer was proposed by other researchers, and so as all the material used in this study (training videos, slides, questionnaires, projects descriptions). Moreover, this bias is reduced since SaPeer and its artifacts were extensively validated by its proponents in previous works. To mitigate *ethical issues*, during the classes, it was reinforced that students were providing feedback regarding an approach available in the literature that was not developed by the professor. Besides, it is not possible to relate students' opinions and their names, all information was analyzed anonymously, and students were not graded based on the content of their feedback, their answers to the questionnaires and interviews, but only on the final documents produced.

Regarding *external validity*, we believe that our results are applicable in similar educational contexts. We compared our results with the original study, but we could not quantitatively compare effectiveness due to our different design. As mentioned, this is mitigated by qualitatively analyzing the students' feedback.

5   Results

RQ1: What are the most frequent mistakes performed by the students with respect to the original study?

We compare the results with the original study, considering the average between self- and peer- reviews for the first interview. This comparison is presented in Fig. 1.

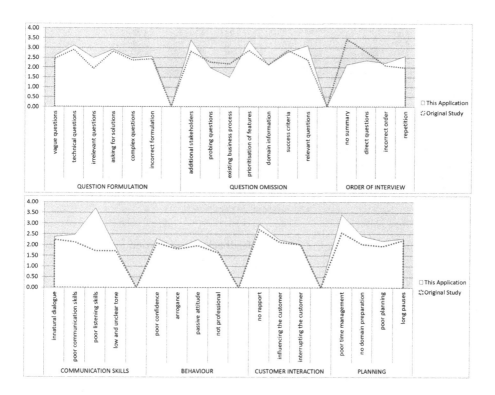

Fig. 1. Comparison between the results of the application and the original study.

We observe a similar behaviour between this work and the original one regarding the mistakes—the plot lines tend to overlap—although the students of this application committed slightly more errors. Common frequent mistakes in both groups are: not asking for additional stakeholders, not asking for prioritization, not asking about success criteria, limited rapport, and poor time management.

The categories that present more differences are communication skills, order of interview, and planning. In particular, our students performed worse in terms of listening skills and time management. This may be linked to the limited time they were given in the interview (10 vs 15 min), suggesting that the time constraint given is probably too limited. More time should be allocated to reduce the hurry that may lead to asking one question after the other without listening for the answer. On the other hand, our students remembered to perform a summary at the end of the interview. Different from the original study, the tutor provided this suggestion already before the first interview. This suggests that some mistakes can be actually corrected easily when the students know them, and this can be particularly useful to address the relevant mistakes in question omission mentioned above.

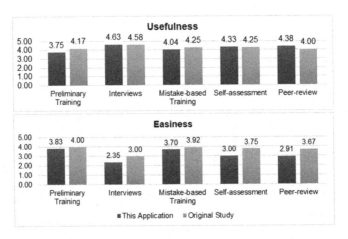

Fig. 2. Comparison between the students' feedback of application and original study.

RQ2: What is the degree of easiness and usefulness of the approach from the viewpoint of the students?

In Fig. 2, we present the results about usefulness and easiness compared with the original study. Values are in general, very close, but some nuanced differences can be noticed. We observe that the students of this application considered the activities of interviews, self-assessment, and especially peer-review more useful than the students from the original study. We believe that these results may be related to the *novelty* of this experience by the students. The training, instead, were considered less useful, probably due to some concerns about the quality of the videos (see RQ3). Concerning easiness, we notice that the averages of the original study are higher than this application in terms of easiness. The limited time given for each task and the quality of the training may have played a role. As in the original study, the activity the students found most difficult was the interviews.

In the feedback questionnaire, some students highlighted the usefulness of the interviews:

"All tasks were very useful, well thought out, and had clear and relevant objectives. But I especially liked the interviews: we were responsible for obtaining all the necessary information for the rest of the project in a few minutes of contact with the client, and for that, we had to use the techniques and tips shown in class to help us to plan."

"All of them were very useful, but in my opinion, the task related to the interview was fundamental, as it was something I had no idea of the importance of this moment for the development of a quality project. In short: aligning our thoughts with the customer is essential."

RQ3: What are the benefits and challenges of SaPeer from the viewpoint of the students?

We organize our discussion by highlighting our observations from applying the approach considering the following categories: challenges, benefits, and improvements as presented in Fig. 3.

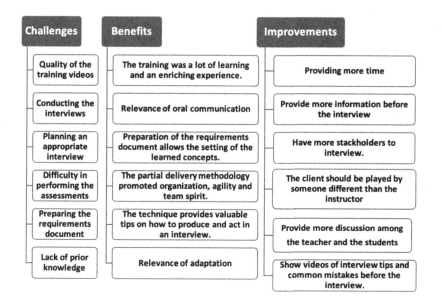

Fig. 3. Results of thematic analysis.

Challenges. Many students reported challenges about the **Quality of the training videos**, and stated that the quality of audio of the videos provided by SaPeer approach should improve. Students performed comments about this: *"The videos have a very bad audio"*; *"I felt a little difficult to understand some parts of the first video (interview tips) due to the audio quality and the absence of subtitles"*; and *"The accent of the person in the videos makes it difficult to understand."*

Some students also highlighted the difficulty in **Conducting the Interviews**: *"The interviews, especially the second, where you must control yourself so as not to repeat the mistakes made in the first. On how to express yourself and what questions to ask in that reduced time frame, in order to extract as much information from the stakeholder"*. Difficulties in **Planning an appropriate interview** observed in RQ1 are also confirmed by the students: **"An error was the lack of elaboration of a plan with other questions depending on the direction of the interview"**. Other themes are concerned with difficulties encountered in the other activities, namely assessment and requirements specification. Finally, the **Lack of prior knowledge**, both about the specific interview context and about the discipline and the software engineering process in general, was regarded as a relevant pain point: *"I believe it would be interesting to present some information about the profile of the person interviewed"*; *"The little knowledge about software engineering is one of the most difficult parts"*.

Benefits. The students reported some benefits and contributions regarding the course in general, the training as well as about the application of SaPeer as presented in Fig. 3. The students observed that the **Training provided a lot of learning and an enriching experience**. Some comments are: *"Training was carried out satisfactorily"*, *"It improves the proximity between teacher and students"*; *"I really liked the video about common mistakes when doing interviews with stakeholders, I think that more than self-assessment, the video made me reflect what was missing, what I got right and what I need to learn for an elicitation interview"*; *"I see Interviews as the most useful task, as we experience in a practical way the difficulties that go from planning to understanding the points covered in the interview, in addition to leaving us much better prepared for any future interviews"*.

Another positive feedback was that **Adopting corrective feedback learning approach and asking students to deliver small artifacts contributes to better learning.** A Student pointed out the benefit of such choice: *"I would like to highlight that the partial delivery methodology of the requirements document promoted organization, agility and team spirit in the development of the requirements document"*. Benefits experienced are also concerned with the understanding of the **Relevance of oral communication**: *"Self-assessment of the interview made me face my own mistakes when making a presentation, highlighting the need to prepare myself better for other similar situations, especially considering the importance of good performance in moments of oral communication"*. Also, **Relevance of adaptation** was recognized as something that was learned thanks to the followed approach, as the need for introducing and inventing questions that would go beyond the script was triggered by the improvised interviews: *"It would be interesting for the group to rethink itself during the interview, reserving a period of unplanned questions to avoid addressing questions already answered"*; *"I believe that trying to deal with the path that the interview took because in some moments it leaves the script that we planned"*.

Improvements. The students also provided some suggestions to improve the course. The most reported improvement was that **More time is necessary.** In this work, the course was distributed in five days in an intensive week where each class had three hours of duration per day. Although SaPeer recommends that students should be given around 3 weeks to work on all the activities of the module [15], we did not have this time available for the training. This short time was pointed out by the students: *"In the end, the project's task was a little complex with respect to time. But I think we did a good job, even with a little pressure"*; *"Recommendation would be to give more time for delivery of the requirements document, I found the delivery time a little tight"*; *"The scarcity of time in planning and consequently the short time to execute something unplanned compromises a better performance"*. Other recommendations are concerned with the organization of the interview, as students missed the opportunity of interviewing more than one stakeholder: *"A suggestion would be to have one more stakeholder (if possible), so that students have access to another opinion about the problem"*.

Students also had some concerns about the tutor and noticed that **The person playing the role of customer needs to better prepare himself/herself to the interviews**: *"The customer should provide the same information for all groups in the same project"*. We actually observed that sometimes we forgot what was said to one group, and this was noticed by the students thanks to the peer-review questionnaire. This is not a major issue, as small imbalances in information should not affect the learning outcomes. However, it may decrease the trust of the students towards the preparation of the teacher. To mitigate this problem, the professor started to take notes of the answers provided. Maybe having other tutors might help. However, this should be done carefully in order to maintain consistency between information shared by all tutors.

Other students suggested that **The role of the client should be played by someone different than the instructor**: *"The students themselves could play the role of customer"*; *"The customer could be from the other group"*. We recommend that someone else plays this role, such as a tutor, since having the same person conducting the classes, playing the role of customer and assigning grades may confuse or inhibits some students because of the feeling of evaluation. This was not possible in our work because there were no tutors available for the classes, and the instructor had to perform all tasks himself/herself. The students suggested that other students may play the role of customer. Although this also the approach adopted by previous works [4,5], and we considered adopting this practice during the classes preparation, it requires a great effort of coordination planning, monitoring and training the students to behave as clients. Hence, it would be a huge effort to be done in a discipline that lasts only one week.

A final common recommendation is **Show videos of interview tips and common mistakes before the interview**: *"I think the first interview would be more complete if we knew what the best practices for interviews are and what mistakes we should avoid making"*. As we noticed, the recommendations concerning the need to perform a summary were actually provided beforehand, and this error was generally avoided. Although in principle SaPeer suggests learning from committed mistakes, we argue that in case of limited time the training could be performed before the first interview.

6   Conclusions and Further Research

In this paper, we present an external application of the evaluation of the SaPeer approach [12]. We provide the following take-away messages: 1) students tend to commit the same mistakes, especially in the areas of question formulation, behaviour, and customer interaction; 2) students struggle more in the areas of question omission and planning; 3) significant differences are observed for the mistake "no summary" (at the end of the interview), quite common in the original study, and less common here, thanks to a simple recommendation provided at the beginning of the lectures; 4) the steps of SaPeer are confirmed to be useful; 5) interviews are confirmed to be among the most useful steps, but also the most difficult 6) time is a relevant issue, and a training that gives more time for interviews, while sacrificing aspects that are considered less useful (e.g., preliminary training or peer-review), can be an appropriate direction in contexts with limited resources. The contributions of this work are:

- We confirmed and provided evidence that the material of the approach is reusable, it is possible to adapt and replicate it.
- We provide an external application in a lower resources context (one professor to do all interviews and restrictions of time—one week only), supporting the empirical grounding of REET research.
- We had qualitative indications that, in this study, we obtained learning outcomes similar to the original. We observed that the ability of the students to conduct requirements elicitation interviews was improved, as well as their ability to analyze the execution and the content of requirements elicitation interviews. We conclude this by analyzing students' feedback (Sect. 5, RQ3), and taking into account their overall performance (meeting notes and requirements specification documents produced).
- We provide additional information through observations, lessons learned and suggestions for improvements, especially: improve video lectures, possibly adapting to the language of the students; have a tutor in the role of customer, instead of the instructor; have different stakeholders in different interviews.

As future works, we expect to join forces with the SaPeer team to carry our further applications of the approach, also including *role reversal* [13], and especially allowing more time for interviews. The SaPeer team involves tutors from the US and Australia, and the integration of a Brazilian viewpoint can allow the REET community to better understand the *cultural* differences that can emerge in applying the approach. These differences may not be clearly identifiable with an external application, such as the current one, in which culture is not considered a primary viewpoint for comparison.

Acknowledgements. Authors would like to thank all the students who participated in this study.

References

1. Abrahão, S., Insfran, E., Carsí, J.A., Genero, M.: Evaluating requirements modeling methods based on user perceptions: a family of experiments. Inf. Sci. **181**(16), 3356–3378 (2011)
2. Baldassarre, M.T., Carver, J., Dieste, O., Juristo, N.: Replication types: towards a shared taxonomy. In: EASE 2014, pp. 1–4 (2014)
3. Bano, M., Zowghi, D., Ferrari, A., Spoletini, P.: Inspectors academy: pedagogical design for requirements inspection training. In: IEEE RE 2020, pp. 215–226. IEEE (2020)
4. Bano, M., Zowghi, D., Ferrari, A., Spoletini, P., Donati, B.: Learning from mistakes: an empirical study of elicitation interviews performed by novices. In: IEEE RE 2018, pp. 182–193. IEEE (2018)
5. Bano, M., Zowghi, D., Ferrari, A., Spoletini, P., Donati, B.: Teaching requirements elicitation interviews: an empirical study of learning from mistakes. Requirements Eng. **24**(3), 259–289 (2019). https://doi.org/10.1007/s00766-019-00313-0
6. Carver, J.C., Juristo, N., Baldassarre, M.T., Vegas, S.: Replications of software engineering experiments. EMSE **19**, 267–276 (2014). https://doi.org/10.1007/s10664-013-9290-8
7. Coughlan, J., Macredie, R.D.: Effective communication in requirements elicitation: a comparison of methodologies. Requirements Eng. **7**(2), 47–60 (2002). https://doi.org/10.1007/s007660200004
8. Davis, A., Dieste, O., Hickey, A., Juristo, N., Moreno, A.M.: Effectiveness of requirements elicitation techniques: empirical results derived from a systematic review. In: IEEE RE 2006, pp. 179–188. IEEE (2006)
9. De Ascaniis, S., Cantoni, L., Sutinen, E., Talling, R.: A lifelike experience to train user requirements elicitation skills. In: Marcus, A., Wang, W. (eds.) DUXU 2017. LNCS, vol. 10290, pp. 219–237. Springer, Cham (2017). https://doi.org/10.1007/978-3-319-58640-3_16
10. Donati, B., Ferrari, A., Spoletini, P., Gnesi, S.: Common mistakes of student analysts in requirements elicitation interviews. In: Grünbacher, P., Perini, A. (eds.) REFSQ 2017. LNCS, vol. 10153, pp. 148–164. Springer, Cham (2017). https://doi.org/10.1007/978-3-319-54045-0_11
11. Fernández, D.M., et al.: Naming the pain in requirements engineering. Empir. Softw. Eng. **22**(5), 2298–2338 (2016). https://doi.org/10.1007/s10664-016-9451-7
12. Ferrari, A., Spoletini, P., Bano, M., Zowghi, D.: Learning requirements elicitation interviews with role-playing, self-assessment and peer-review. In: IEEE RE 2019, pp. 28–39. IEEE (2019)
13. Ferrari, A., Spoletini, P., Bano, M., Zowghi, D.: Sapeer and reversesapeer: teaching requirements elicitation interviews with role-playing and role reversal. Requirements Eng. **25**, 1–22 (2020). https://doi.org/10.1007/s00766-020-00334-0
14. Ferrari, A., Spoletini, P., Gnesi, S.: Ambiguity and tacit knowledge in requirements elicitation interviews. Requirements Eng. **21**(3), 333–355 (2016). https://doi.org/10.1007/s00766-016-0249-3
15. Ferrari, A., Spoletini, P., Bano, M., Zowghi, D.: Sapeer approach for training students in requirements elicitation interviews–educational material (2020). https://zenodo.org/record/3765214

16. Gabrysiak, G., Giese, H., Seibel, A., Neumann, S.: Teaching requirements engineering with virtual stakeholders without software engineering knowledge. In: REET 2010, pp. 36–45. IEEE (2010)

17. Garcia, I., Pacheco, C., Méndez, F., Calvo-Manzano, J.A.: The effects of game-based learning in the acquisition of "soft skills" on undergraduate software engineering courses: a systematic literature review. Comput. Appl. Eng. Educ. **28**(5), 1327–1354 (2020)

18. Hadar, I., Reinhartz-Berger, I., Kuflik, T., Perini, A., Ricca, F., Susi, A.: Comparing the comprehensibility of requirements models expressed in use case and tropos: results from a family of experiments. Inf. Softw. Technol. **55**(10), 1823–1843 (2013)

19. Hadar, I., Soffer, P., Kenzi, K.: The role of domain knowledge in requirements elicitation via interviews: an exploratory study. Requirements Eng. **19**(2), 143–159 (2012). https://doi.org/10.1007/s00766-012-0163-2

20. Hertz, K., Spoletini, P.: Are requirements engineering courses covering what industry needs? a preliminary analysis of the United States situation. In: REET 2018, pp. 20–23. IEEE (2018)

21. Hu, W., Carver, J.C., Anu, V., Walia, G.S., Bradshaw, G.L.: Using human error information for error prevention. Empir. Softw. Eng. **23**(6), 3768–3800 (2018). https://doi.org/10.1007/s10664-018-9623-8

22. Khatwani, C., Jin, X., Niu, N., Koshoffer, A., Newman, L., Savolainen, J.: Advancing viewpoint merging in requirements engineering: a theoretical replication and explanatory study. Requirements Eng. **22**(3), 317–338 (2017). https://doi.org/10.1007/s00766-017-0271-0

23. Nakamura, T., Kai, U., Tachikawa, Y.: Requirements engineering education using expert system and role-play training. In: IEEE TALE 2014, pp. 375–382. IEEE (2014)

24. Nkamaura, T., Tachikawa, Y.: Requirements engineering education using role-play training. In: IEEE TALE 2016, pp. 231–238. IEEE (2016)

25. Ouhbi, S.: Evaluating role playing efficiency to teach requirements engineering. In: 2019 IEEE Global Engineering Education Conference (EDUCON), pp. 1007–1010. IEEE (2019)

26. Ouhbi, S., Idri, A., Fernández-Alemán, J.L., Toval, A.: Requirements engineering education: a systematic mapping study. Requirements Eng. **20**(2), 119–138 (2013). https://doi.org/10.1007/s00766-013-0192-5

27. Ouhbi, S., Pombo, N.: Software engineering education: challenges and perspectives. In: 2020 IEEE Global Engineering Education Conference (EDUCON), pp. 202–209. IEEE (2020)

28. Regev, G., Gause, D.C., Wegmann, A.: Requirements engineering education in the 21st century, an experiential learning approach. In: IEEE RE 2008, pp. 85–94. IEEE (2008)

29. Rueda, S., Panach, J.I., Distante, D.: Requirements elicitation methods based on interviews in comparison: a family of experiments. Inf. Softw. Technol. **126**, 106361 (2020)

30. Spoletini, P., Ferrari, A., Bano, M., Zowghi, D., Gnesi, S.: Interview review: an empirical study on detecting ambiguities in requirements elicitation interviews. In: Kamsties, E., Horkoff, J., Dalpiaz, F. (eds.) REFSQ 2018. LNCS, vol. 10753, pp. 101–118. Springer, Cham (2018). https://doi.org/10.1007/978-3-319-77243-1_7

31. Svensson, R.B., Regnell, B.: Is role playing in requirements engineering education increasing learning outcome? Requirements Eng. **22**(4), 475–489 (2017). https://doi.org/10.1007/s00766-016-0248-4

32. Vilela, J., Lopes, J.: Evaluating the students' experience with a requirements elicitation and communication game. In: Conferencia Iberoamericana de Software Engineering (CIBSE). CIBSE (2020)

33. Walia, G.S., Carver, J.C.: Using error abstraction and classification to improve requirement quality: conclusions from a family of four empirical studies. Empir. Softw. Eng. **18**(4), 625–658 (2013). https://doi.org/10.1007/s10664-012-9202-3

34. Wohlin, C., Runeson, P., Höst, M., Ohlsson, M.C., Regnell, B., Wesslén, A.: Experimentation in Software Engineering. Springer, Heidelberg (2012). https://doi.org/10.1007/978-3-642-29044-2

35. Zowghi, D., Paryani, S.: Teaching requirements engineering through role playing: Lessons learnt. In: IEEE RE 2003, pp. 233–241. IEEE (2003)

On Understanding the Relation of Knowledge and Confidence to Requirements Quality

Razieh Dehghani[1] (ID), Krzysztof Wnuk[2] (ID), Daniel Mendez[2,3] (ID), Tony Gorschek[2] (ID), and Raman Ramsin[1](✉) (ID)

[1] Department of Computer Engineering, Sharif University of Technology, Tehran, Iran
`rdehghani@ce.sharif.edu, ramsin@sharif.edu`
[2] Department of Software Engineering, SERL Sweden/Blekinge Institute of Technology, Karlskrona, Sweden
`{Krzysztof.wnuk,daniel.mendez,tony.gorschek}@bth.se`
[3] Fortiss GmbH, Munich, Germany

Abstract. **[Context and Motivation]** Software requirements are affected by the knowledge and confidence of software engineers. Analyzing the interrelated impact of these factors is difficult because of the challenges of assessing knowledge and confidence. **[Question/Problem]** This research aims to draw attention to the need for considering the interrelated effects of confidence and knowledge on requirements quality, which has not been addressed by previous publications. **[Principal ideas/results]** For this purpose, the following steps have been taken: 1) requirements quality was defined based on the instructions provided by the ISO29148:2011 standard, 2) we selected the symptoms of low qualified requirements based on ISO29148:2011, 3) we analyzed five Software Requirements Specification (SRS) documents to find these symptoms, 3) people who have prepared the documents were categorized in four classes to specify the more/less knowledge and confidence they have regarding the symptoms, and 4) finally, the relation of lack of enough knowledge and confidence to symptoms of low quality was investigated. The results revealed that the simultaneous deficiency of confidence and knowledge has more negative effects in comparison with a deficiency of knowledge or confidence. **[Contribution]** In brief, this study has achieved these results: 1) the realization that a combined lack of knowledge and confidence has a larger effect on requirements quality than only one of the two factors, 2) the relation between low qualified requirements and requirements engineers' needs for knowledge and confidence, and 3) variety of requirements engineers' needs for knowledge based on their abilities to make discriminative and consistent decisions.

Keywords: Requirements quality · Requirements engineers' confidence · Requirements engineering · Requirements engineering knowledge

1 Introduction

Building software solutions requires achieving sufficient requirements quality. Requirements quality is affected by humans, processes and tools [1]. Requirements engineers'

© Springer Nature Switzerland AG 2021
F. Dalpiaz and P. Spoletini (Eds.): REFSQ 2021, LNCS 12685, pp. 208–224, 2021.
https://doi.org/10.1007/978-3-030-73128-1_15

knowledge and their confidence are the two human-related factors that affect requirements quality. Researchers have previously assessed the effect of these factors separately. For example, it has been found that the effectiveness of interviews is affected by domain knowledge [2, 3]. Also, the relation between engineers' confidence and some specific types of requirements, such as safety, has been investigated [4].

Figure 1 shows the research model used in this study. Hypotheses H4 and H5 refer to the effects of requirements engineers' knowledge and confidence on requirements quality. Since knowledge and confidence are interrelated [5], this research has focused on assessing the effects of knowledge deficit and a lack of confidence (hypotheses H6 and H7). Other relations, shown in Fig. 1, refer to the methods that have been used for assessing quality, knowledge, and confidence, as follows:

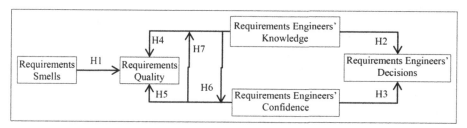

Fig. 1. Research model

1) *Requirements Quality*: The ISO29148:2011 standard has provided a set of detailed principles for producing qualified SRS documents [6]. On this basis, Femmer et al. have defined the term *Requirements Smell* to assess quality [7]. Requirements smell is *"an indicator of a quality violation, which may lead to a defect, with a concrete location and a concrete detection mechanism"* [7]. Smells help find the location for low-qualified requirements. The location refers to the word/sentence, which violates the quality. For example, a vague adjective is a location for a low-qualified requirement because it might result in a misunderstanding about the requirement. It should be noted that the location might vary based on the product in which the requirements are stored. We have focused on SRS documents and used smells for assessing the quality of requirements (H1 in Fig. 1).

2) *Requirements Engineers' Knowledge*: This term is defined from a capability-based perspective. From this point of view, knowledge is "the potential to influence action" [8]. On this basis, *requirements engineers' knowledge* has the potential to influence the process of preparing SRS documents. Assessing the time that an individual spends in requirements engineering, namely experience, is a method for assessing knowledge. Besides, defects in decisions made by requirements engineers are symptoms of their level of expertise. In this research, low experience and inability to make discriminative and consistent decisions are considered as the symptoms of lack of enough knowledge [9, 10]. Discrimination and consistency have been defined from a comparative point of view [9]. On this basis, compared to novices, experts make

more consistent and discriminative decisions throughout the requirements engineering process. Thus, H2 shows that inability in making discriminative and consistent decisions was used as the symptom for lack of enough knowledge.

3) *Requirements Engineers' Confidence*: This term refers to the feeling of trust about the SRS document that is prepared/reviewed/used. On this basis, uncertainty in making requirements engineering decisions was chosen as the symptom for low confidence (H3 in Fig. 1). This has been inspired by the results of Boness et al.'s research [11]. They have defined this term by proposing four criteria for refuting/warranting a claim about *requirements engineers' confidence* in goal-oriented requirements analysis. On this basis, we have proposed the following measures to assess confidence regarding various dimensions of requirements smells: depth of coverage, breadth of coverage, correctness, achievability, assumption, and accuracy. Analyzing the data about these criteria helps refute/warrant our claim about requirements engineers' certainty. It should be noted that uncertainty might occur regarding various features of requirements. We cannot claim that our research covers all these dimensions. However, by studying the research that has previously been conducted, we have tried to choose some specific dimensions of uncertainty regarding each dimension of requirements smells.

It should be noted that the methods we have used for assessing knowledge, confidence, and quality are context-independent [1, 9, 11]. However, some factors might affect the assessment. For example, the cultural features might affect requirements engineers' decisions [12].

The novelty of this research comes by addressing three issues: (1) in previous related work, requirements smells have not been traced so far to requirements engineers' knowledge and confidence, (2) abilities in making decisions have not been considered as symptoms of lack of requirements engineers' knowledge, and (3) interrelations between knowledge and confidence have not yet been considered.

Addressing these issues is important because: (1) requirements smells help trace the effect of low confidence and/or knowledge to a specific location(s) for low qualified requirements, (2) experience in requirements engineering, which refers to the time spent in academia and industry for requirements engineering, is not the only factor which affects individuals' knowledge; thus this research considers the skills in making requirements engineering decisions as well, and (3) ignoring the effect of low confidence or low knowledge yields wrong results and thus leads to inability to eliminate the causes for low quality.

The rest of this paper is organized as follows: next section provides an overview of the work related to this research; then, the method for conducting research and collecting data is explained; thereafter, results of analyzing the data are provided; and finally, the paper is ended by providing the conclusions and also suggesting some ways to further this research.

2   Background and Related Work

This work focuses on the intersection of three concepts: requirements quality, requirements engineers' knowledge and confidence. Femmer et al. have introduced requirements smells to assess the quality of SRS documents [7]. Similarly, Shanteau et al. and

Boness et al. have respectively analyzed the individuals' knowledge and confidence by scrutinizing the decisions they make [9, 11]. Figure 2 shows the terminologies and the relationships between the main concepts used in this work.

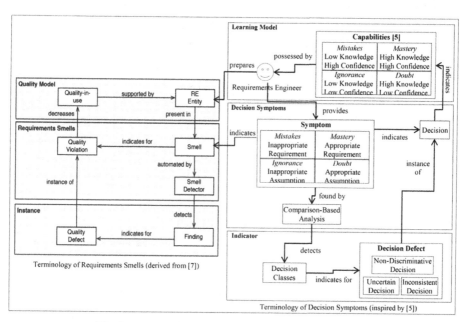

Fig. 2. Terminology of the concepts used in this research

The left part of Fig. 2 is derived from Femmer et al.'s study [7] and presents the terminology for requirements smells. As explained in the first section, the requirements that do not follow the instructions provided by ISO29148:2011 standard [6], namely requirements smells, are low qualified [7]. We have categorized the research in the area of effects of smells as follows [13]:

- Effects of smells on *artifacts*: This category is concerned about the effects of smells on artifacts produced throughout the software development process. SRS is an example of an artifact affected by defects of natural languages [7].
- Effects of smells on *processes*: Research in this area addresses the effects of smells on development processes. As an example, the effect of requirements smells on test case design has been discussed in [14].
- Effects of smells on *people*: This area of research has been addressed indirectly. For example, Bjarnason et al. have provided a schema of requirements flow to depict the effect of requirements change on developers and customers [15].

Table 1 provides a high-level overview of various categories of smells. The source of the smells is provided in the third column. The smells might be related to requirements, the requirements process, or the time/place/logic/people-dependent conditions and constraints. It should be noted that the measures have been proposed based on the issue that is emphasized within the reference from which it has been elicited. More measures might also be elicited by other researchers.

Table 1. Categories of requirements smells

Smell dimension	Smell category	Measure [reference] (ID-S#)
Requirement	Ambiguity	Probability of various interpretations regarding the meaning of requirement [7] (ID-S1)
	Incompleteness	Probability of having non-elicited requirements [16] (ID-S2)
	Inconsistency	Probability of having inconsistent requirements [16] (ID-S3)
	Redundancy	Probability of having redundant requirements [16] (ID-S4)
	Incorrectness	Probability of having semantically incorrect requirements [16] (ID-S5)
	Size	Probability of having compound requirements [16] (ID-S6)
		Probability of having large SRS documents [16] (ID-S7)
RE process	Analysis	Probability of having an inappropriate data collection method [16] (ID-S8)
		Probability of having non-identified stakeholders [6] (ID-S9)
		Probability of wrong judgment about criticality and risks [6] (ID-S10)
	Documentation	Probability of lack of explanation about "domain-specific and frequently occurring concepts" [16] (ID-S11)
		Probability of having an incomplete glossary [16] (ID-S12)
	Verification	Probability of having inappropriate requirements verification method [16] (ID-S15)
	Validation	Probability of having requirements, non-traceable to stakeholders [16] (ID-S14)
		Probability of having non-defined "stakeholder requirements for validation" [16] (ID-S13)
	Management	Probability of having products non-traceable to requirements [6] (ID-S16)
		Probability of having quality requirements without measures [6] (ID-S17)

(continued)

Table 1. (*continued*)

Smell dimension	Smell category	Measure [reference] (ID-S#)
Time-dependent conditions and constraints	Ambiguity	Probability of uncertainty about the order for satisfying the requirements [16] (ID-S18)
		Probability of uncertainty about time for verification [6] (ID-S19)
	Incompleteness	Probability of having missing time-dependent conditions and constraints [16] (ID-S20)
Place-dependent conditions and constraints	Ambiguity	Probability of having functionalities outside the boundaries of software architecture [6] (ID-S21)
		Probability of making mistakes regarding system boundary [6] (ID-S22)
		Probability of misalignment between stakeholder, system, and software requirements [6] (ID-S23)
		Probability of having ambiguous "venue and environment for verification" [6] (ID-S24)
	Incompleteness	Probability of having unrecognized external elements (including regulations, culture, etc.) [6] (ID-S25)
		Probability of having an incomplete configuration baseline [6] (ID-S26)
		Probability of missing the constraints that affect the architecture [6] (ID-S27)
	Unavailability	Probability of inability in obtaining "items of information" [6] (ID-S28)
People-dependent conditions and constraints	Ambiguity	Probability of uncertainty about stakeholders' preferences [16] (ID-S37)
		Probability of uncertainty about interactions between users and systems [16] (ID-S38)
	Inconsistency	Probability of having wrong priorities regarding inconsistent stakeholders' requirements [16] (ID-S39)
	Incompleteness	Probability of specifying wrong individuals for conducting verification [16] (ID-S40)
		Probability of having wrong supportive information about stakeholders [6] (ID-S41)

(*continued*)

Table 1. (*continued*)

Smell dimension	Smell category	Measure [reference] (ID-S#)
Logic-dependent conditions and constraints	Ambiguity	Probability of unavailability of metadata regarding requirements [16] (ID-S29)
		Probability of having open-ended sentences [16] (ID-S30)
		Probability of having vague dependencies between requirements [16] (ID-S31)
		Probability of having wrong overall integrity of requirements [6] (ID-S32)
		Probability of having wrong estimations regarding goal satisfaction [16] (ID-S33)
		Probability of having vague control flows [16] (ID-S34)
		Probability of having vague logic behind optional requirements [16] (ID-S35)
	Incompleteness	Probability of having non-maintained rationale and assumptions [6] (ID-S36)

The right part of Fig. 2 presents the terminology for decision symptoms. The "possessed by" arrow shows that each requirements engineer has some capabilities. Defects in making requirements engineering decisions are considered as symptoms of a lack of knowledge (including experience) or confidence. The defects can be classified as follows: 1) inappropriate assumptions are symptoms of ignorance because of low knowledge and confidence, 2) appropriate requirements indicate mastery in RE due to a high level of knowledge and confidence, 3) inappropriate requirements are symptoms of making mistakes because of low knowledge and high confidence, and 4) appropriate assumptions indicate doubt in RE due to high knowledge and low confidence.

It should be noted that the decisions might be inappropriate due to various reasons. That is why we have added a "decision classes" component in Fig. 2. As mentioned, we have selected three instances of defects, which are symptoms of inappropriateness, as follows:

1) *Uncertainty* is an indicator of the need for more confidence. Figure 3 shows the model for assessing confidence. This is inspired by the procedures used in courts to refute/warrant a claim [17]. This method has previously been used for assessing confidence in requirements analysis, as well [11]. As shown, we first claim that the requirements engineer is not confident. Then, we look for the reasons through which we can warrant or refute our claim. To find the warranting and violating reasons, we have used the results of Boness et al.'s research (Table 2). As shown in Table 2, some measures have been proposed for assessing confidence regarding various

dimensions of smells. It should be noted that these measures are not the complete set, and more measures might be added by researchers.

2) Inability to make *consistent* and *discriminative* decisions is the symptom of a lack of knowledge. It should be noted that experience is also a helpful factor for providing some assumptions about someone's knowledge, though it is not an accurate measure. We thus judged these assumptions by analyzing the decisions by using the CWS ratio [9, 10].

Table 2. Confidence factors (Inspired by [11])

Smell dimension (Smell in)	Confidence dimension	Measure (Confidence Factor) (ID-C#)
Requirement	What	Depth of Coverage: Confidence that the requirements have been adequately scrutinized in-depth (similar to refinement [11]) (ID-C1)
		Breadth of Coverage: Confidence that the requirements have been adequately scrutinized in breadth (similar to engagement [11]) (ID-C2)
		Correctness: Confidence that the requirements are correct (ID-C3)
		Achievability [11]: Confidence that the requirements are achievable (ID-C4)
RE process	How	Depth of Coverage: Confidence that the RE process has adequately covered the fine-grained RE tasks (ID-C5)
		Breadth of Coverage: Confidence that the RE process has adequately covered the general RE process (ID-C6)
		Correctness: Confidence that the RE process has been performed in the right way (ID-C7)
Time-dependent conditions and constraints	When	Achievability: Confidence that the time-dependent conditions are achievable [11] (ID-C8)
		Assumption: Confidence that the time-dependent constraints are sound [11] (ID-C9)
		Accuracy: Confidence that the time-dependent conditions are specified (ID-C16)
Place-dependent conditions and constraints	Where	Achievability: Confidence that the place-dependent conditions are achievable [11] (ID-C10)
		Assumption: Confidence that the place-dependent constraints are sound [11] (ID-C11)
		Accuracy: Confidence that the place-dependent conditions are specified (ID-17)

(continued)

Table 2. (*continued*)

Smell dimension (Smell in)	Confidence dimension	Measure (Confidence Factor) (ID-C#)
Logic-dependent conditions and constraints	Why	Achievability: Confidence that the logic-dependent conditions are achievable [11] (ID-C12)
		Assumption: Confidence that the logic-dependent constraints are sound [11] (ID-C13)
		Accuracy: Confidence that the logic-dependent conditions are specified (ID-C18)
People-dependent conditions and constraints	Who	Achievability: Confidence that the people-dependent conditions are achievable [11] (ID-C14)
		Assumption: Confidence that the people-dependent constraints are sound [11] (ID-C15)
		Accuracy: Confidence that the people-dependent conditions are specified (ID-C19)

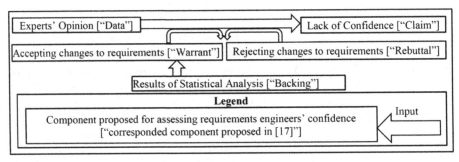

Fig. 3. Model for confidence assessment (Derived from [17])

3 Research Methodology and Data Collection

Figure 4 presents the research steps followed in this work. First, we analyzed five SRS documents prepared by graduate students at the Blekinge Institute of Technology (BTH) in the course of their project work in Requirements Engineering and identified requirements smells in these documents. "The database should be reliable" is an example of a vague sentence (requirements smell). Next, we analyzed the project grading criteria to get aware of the requirements necessary for preparing the SRS documents and, thus, could not be considered as smells. Next, we designed the questionnaires to analyze students' knowledge and confidence, inspired by the Smith et al.'s four quadrants based on the level of human knowledge and confidence [5]: "Ignorance" (low knowledge and low confidence), "Doubt" (low confidence and high knowledge), "Mistakes" (high confidence and low knowledge), and "Mastery" (high confidence and high knowledge).

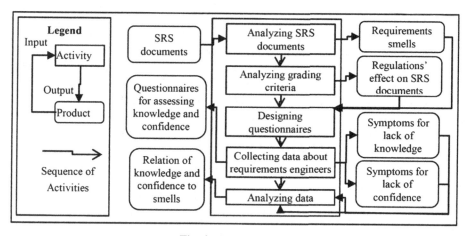

Fig. 4. Research steps

The questions were answered by students who have prepared the SRS documents. Thus, the questionnaires encompass questions regarding certainty about specific smells found in the SRS documents, and students' abilities to make discriminative and consistent decisions. Examples of the questions are provided at the end of the paper, in the Appendix section. As an example, the students who have elicited the requirement about reliability *doubt* about the criteria by which reliability would be assessed. It should be noted that with the aim of alleviating the effect of environmental factors that might affect students' responses, the professors assured the students that the responses would not affect the grades.

Thus, as explained we have collected data in two steps:

1) *Analyzing SRS documents*: To find the smells, documents were analyzed by using the measures provided in Table 1. The results revealed a list of specific smells within each SRS document.

2) *Analyzing knowledge and confidence*: To assess students' confidence and knowledge regarding requirements, specific questions were designed for each group of students who have prepared the SRS documents. An instance of the instrument (questionnaire) we have designed is provided in the appendix section of this paper. After analyzing the data about students' knowledge and confidence, we could categorize the responses within the four mentioned quadrants. The method for analyzing knowledge and confidence is explained in the following paragraphs. First, we categorized the students based on their knowledge, and then we categorized their responses based on the response that shows the students' certainty/uncertainty.

To warrant or refute our claim about students' confidence, respondents were suggested to apply some changes to their documents, and they could "Agree" or "Disagree" with our suggestions. The changes were suggested in relation to the smells we have found in the first step. As shown in Fig. 3, agreeing with applying the changes was considered as a reason for warranting our claim about the lack of confidence. On the contrary, disagreeing with applying changes was a reason for rebutting our claim.

CWS ratio (Formula 1) was used [9, 10] for assessing the knowledge level. The abbreviation "CWS" comes from the names of individuals who have proposed it. This abbreviation "is used to establish that someone behaves more (high value) or less (low value) as an expert" [9]. In other words, this metric claims that judgments made by experts, in comparison with judgments made by novices, are more discriminating and consistent. According to [9], as shown in Table 3, for diagnostic decisions, there is a greater difference between decisions made by experts and novices, while for non-diagnostic ones, decisions are more similar.

$$\text{"CWS=Discrimination/Inconsistency" [9]} \qquad (1)$$

Table 3. Difference between CWS ratios (Derived from [9])

	Important (Diagnostic)	Partially important (partially-diagnostic)	Non-important (non-diagnostic)
Experts	A	B	C
Novices	D	E	F

Difference between CWS ratio for experts and novices: A-D > B-E > C-F

We have calculated the discrimination and inconsistency factors, provided in Formula 1, as follows: 1) first, we have investigated the number of years that respondents have experienced RE, and thus conducted a preliminary categorization regarding the respondents' expertise; 2) then, we have provided three categories of sentences, and respondents were asked to categorize them within the following classes: "Diagnostic" (important for eliciting the requirements), "Partially-diagnostic" (partially important for eliciting the requirements), and "Non-diagnostic" (not important for eliciting the requirements); 3) after that, we have measured the "Inconsistency" metric by calculating the "average of within-cell variances" ("low variance implies high consistency") [9]; 4) thereafter, the "Discrimination" metric was obtained by calculating mean square values ("High variance implies high discrimination") [9]; 5) after that, to calculate the CWS ratio (Formula 1), the discrimination metric was divided by the inconsistency metric; and 6) finally, we have reassessed our judgments about respondents' expertise by moving the students within categorizations so that we could make sure that experts are better in making consistent and discriminative decisions.

4   Results of Data Analysis

Eight groups of students (thirty-three individuals) participated in this study; however, we had to ignore the responses provided by three groups because more than half of the members of these groups did not fill in the questionnaires. The following paragraphs respectively discuss the results of analyzing the data collected for finding the relation of smells to confidence, knowledge, and both knowledge and confidence.

1) *Analyzing data about confidence*: Table 4 provides an example of the responses we have received to assess confidence; rows represent question numbers, and columns represent respondent numbers. As shown, the changes suggested for four questions were agreed upon by at least half of the group members. Table 5 shows the number of respondents who agreed with making the changes suggested for each group; rows represent question numbers, and columns represent group numbers. As shown, we found that except for eight changes, all other ones were agreed to be applied by at least half of the respondents. Thus, it is concluded that the students have confirmed that they are not confident regarding the requirements smells we have found.

Table 4. Example of responses to questions for assessing confidence (Group 1)

	R1	R2	R3	R4	TNA
Q1	"0"	"0"	"1"	"1"	2
Q2	"0"	"0"	"0"	"0"	0
Q3	"1"	"1"	"1"	"1"	4
Q4	"1"	"1"	"1"	"1"	4
Q5	"1"	"1"	"0"	"0"	2

Legend: "1" refers to agreeing with applying the change, and "0" refers to disagreeing with applying the change
TNA stands for Total Number of Agreements

Table 5. Number of respondents who agreed with making the changes (Groups 1–5)

	G1	G2	G3	G4	G5
Q1	2	5	6	5	6
Q2	**0**	4	5	3	3
Q3	4	**1**	4	4	5
Q4	4	4	3	5	5
Q5	2	4	5	5	4
Q6	3	5	4	4	3
Q7	**1**	5	4	3	4
Q8	4	5	6	5	6
Q9	3	5	6	4	5
Q10	3	3	5	5	5
Q11	**1**	3	3	3	4
Q12	2	**2**	5	3	4
Q13	4	**1**	6	3	4
Q14	**1**	3	3	3	**2**

Legend: Bold underlined numbers indicate that less than half of the respondents agreed with making the change

2) *Analyzing data about knowledge*: Table 6 provides the results of calculating the CWS ratio. It should be noted that we have calculated this metric by using three pre-classification methods as follows: 1) timespan of experience in academy environments, 2) timespan of experience in non-academy environments, and 3) the total timespan of experience in both academic and non-academic environments. What we found was that for the third type of pre-classification, in comparison with the other two pre-classification methods, the decisions made by experts and novices are more clearly discriminated (as specified in Table 3).

Table 6. CWS Ratio (Pre-classification was made based on the total timespan of experience in both academic and non-academic environments)

Category	Important (Diagnostic)	Partially important (Partially-Diagnostic)	Non-important (Non-Diagnostic)
Group 1			
Experts	4	0.25	0
Novices	0.8	0.13	0
Result: 4–0.8 > 0.25–0.13 > 0–0			
Group 2			
Experts	5	0.34	0
Novices	0.7	0.21	0
Result: 5–0.7 > 0.34–0.21 > 0–0			
Group 3			
Experts	2.5	0.45	0.1
Novices	0.4	0.10	0
Result: 2.5–0.4 > 0.45–0.10 > 0.1–0			
Group 4			
Experts	3	0.24	0.03
Novices	0.7	0.10	0
Result: 3–0.7 > 0.24–0.10 > 0.03–0			
Group 5			
Experts	4.3	0.38	0
Novices	0.6	0.19	0
Result: 4.3–0.6 > 0.38–0.19 > 0–0			

3) *Analyzing data about both knowledge and confidence*: In total, 33 respondents have specified their opinion regarding 14 smells. Thus, we have received 462 (33 multiplied by 14) responses regarding smells. Each of these responses falls into one of the knowledge-confidence quadrants, based on the evaluation made regarding the respondents. As shown in Fig. 5, the "Ignorance" quadrant encompasses the most responses, which means that a combined lack of knowledge and confidence has the most negative effect on requirements smells, and thus requirements quality.

Looking at the results, we draw attention to the following issues:

1) Uncertainty about requirements is an indicator of low-qualified requirements. Practitioners can check requirements engineers' confidence to find requirements that are

Mistakes (Low Knowledge High Confidence) 155 responses	Mastery (High Knowledge High Confidence) 43 responses
Ignorance (Low Knowledge Low Confidence) 172 responses	**Doubt** (High Knowledge Low Confidence) 92 responses

Fig. 5. Number of responses in knowledge-confidence quadrants

potentially low-qualified. Besides, finding and classifying the reasons for uncertainty is an area of research which is needed to be addressed by researchers.

2) Various requirements engineers might elicit different requirements for a unique software system. This is due to the difference in their knowledge. The CWS ratio helps find the differences. Project managers can use this metric to categorize the employee and plan to improve their skills in RE. Besides, researchers should identify the determinative decisions which should be consistent and discriminative.

3) Requirements quality is affected by a collection of factors. Not only the factors but also their relations affect quality. Due to the interrelation between knowledge and confidence, a lack of confidence simultaneously with a lack of knowledge increases the probability of low quality. Researchers should explore such relations, and practitioners should beware of the simultaneous effects of interrelated factors.

5   Conclusions and Future Work

In this research, we explored the relationship between knowledge and confidence, and requirements quality. For this purpose, we have first analyzed five SRS documents developed by the students of Blekinge Institute of Technology. The analysis aimed to find the low qualified requirements, which was done using a set of criteria named requirements smells. In the next step, students' knowledge and confidence were assessed by analyzing their abilities to make discriminative, consistent, and certain decisions. Finally, we have classified smells based on individuals' knowledge and confidence. We found that most smells fall into the class with a lack of confidence and knowledge.

Thus, requirements for smells might be considered as symptoms of a lack of knowledge and/or confidence. Project managers can use this information (the relation between requirements smells, knowledge, and confidence) to find the areas in which some training mechanisms should be used to improve requirements engineers' skills. As an example, they might decide to hold some workshops to improve requirements engineers' skills. The training materials used within these workshops can be decided on this basis.

This research is novel mainly due to considering the interrelation between knowledge and confidence, using a decision-based comparative method for analyzing knowledge, and analyzing requirements quality based on specific symptoms for low quality. However, conducting one experiment in one academic environment is not enough for approving the relations, and more cases should be investigated to approve the results in general. We aim to further this research by conducting more experiments through which we can collect more data.

Acknowledgement. We would like to acknowledge that this work was supported by the KKS foundation through the S.E.R.T. Research Profile project at Blekinge Institute of Technology and the SERL Lab.

Appendix

Some examples of the questions that we have designed are provided herein. The questions in the following three sections are respectively aimed at assessing confidence, analyzing domain knowledge, and investigating knowledge in RE.

Section A: Imagine that a company manager has studied the SRS document that you have prepared for this project, and you are invited to join a team to help develop the system for which you have elicited the requirements. For the first step, the manager provides the following claims about your document and asks you to address them. Please indicate if you agree/disagree?

1) Regarding the following requirement, much more detail is required and still, it should be refined. DL1: "The information shall be presented using HTML5.2 and CSS3 languages." Agree ☐ Disagree ☐
2) More detail about time-dependent conditions and constraints are required for these requirements: PR5: "The web application shall offer the functionality of registration in the web-app.", and PR4: "The web application shall offer the functionality of login in the web-app." Agree ☐ Disagree ☐
3) You are not sure about the appropriate time for verifying the requirements. Agree ☐ Disagree ☐
4) Policy and regulations have been provided. The effects of cultural elements should also be discussed. Agree ☐ Disagree ☐
5) You are not sure about the dependency between some requirements. For example, it seems that some issues regarding the dependency between the following requirements are not explained: PR1: "The web application shall offer the functionality of adding a new movie review.", and PR2: "The web application shall offer the functionality of rating a movie." Agree ☐ Disagree ☐

Section B: Please answer the following questions:

- How many *industrial (non- academic)* projects have you been engaged in to develop a software system, the same as the system you have engineered requirements for (in the role of a project manager, programmer, etc.)?
- How many *academic* projects have you been engaged in to develop a software system, the same as the system you have engineered requirements for (in the role of a project manager, programmer, etc.)?

Please categorize the following issues as important, partially-important, and non-important in selecting the most suitable requirements prioritization techniques.

- Type of requirement (functional/non-functional)
- Support for evaluating requirements
- Caring about requirements dependencies
- Support for coordinating various stakeholders' requirements
- The number of requirements that should be prioritized

Section C: Please answer the following questions:

- How many *industrial* (*non- academic*) projects have you been engaged in for eliciting requirements?
- How many *academic* projects have you been engaged in for eliciting requirements?

Please categorize the following issues as important, partially-important, and non-important for selecting the most suitable requirements elicitation techniques.

- Complementary requirements elicitation techniques that are required to be applied.
- Number of requirements that would be elicited by the technique(s) chosen.
- People-dependent factors (such as culture).
- The time that it would take to elicit the requirements.

References

1. Femmer, H., Vogelsang, A.: Requirements quality is quality in use. IEEE Softw. **36**(3), 83–91 (2018)
2. Aranda, A.M., Dieste, O., Juristo, N.: Effect of domain knowledge on elicitation effectiveness: an internally replicated controlled experiment. IEEE Trans. Softw. Eng. **42**(5), 427–451 (2015)
3. Hadar, I., Soffer, P., Kenzi, K.: The role of domain knowledge in requirements elicitation via interviews: an exploratory study. Requirements Eng. **19**(2), 143–159 (2012). https://doi.org/10.1007/s00766-012-0163-2
4. Ayoub, A., Kim, B., Lee, I., Sokolsky, O.: A systematic approach to justifying sufficient confidence in software safety arguments. In: Ortmeier, F., Daniel, P. (eds.) SAFECOMP 2012. LNCS, vol. 7612, pp. 305–316. Springer, Heidelberg (2012). https://doi.org/10.1007/978-3-642-33678-2_26
5. Smith, C.J., Adams, T.M., Engstrom, P.G., Cushman, M.J., Bruno, J.E.: U.S. Patent No. 8,165,518. Washington, DC: U.S. Patent and Trademark Office (2012)
6. ISO, IEC, IEEE. ISO/IEC/IEEE 29148:2011. https://standards.ieee.org/standard/29148-2011.html. Accessed 06 Nov 2020
7. Femmer, H., Fernández, D.M., Wagner, S., Eder, S.: Rapid quality assurance with requirements smells. J. Syst. Softw. **123**, 190–213 (2017)
8. Alavi, M., Leidner, D.E.: Knowledge management and knowledge management systems: conceptual foundations and research issues. MIS Q. **25**(1), 107–136 (2001)
9. Shanteau, J., Weiss, D.J., Thomas, R.P., Pounds, J.C.: Performance-based assessment of expertise: How to decide if someone is an expert or not. Eur. J. Oper. Res. **136**(2), 253–263 (2002)

10. Hemming, V., Burgman, M.A., Hanea, A.M., McBride, M.F., Wintle, B.C.: A practical guide to structured expert elicitation using the IDEA protocol. Methods Ecol. Evol. **9**(1), 169–180 (2018)

11. Boness, K., Finkelstein, A., Harrison, R.: A method for assessing confidence in requirements analysis. Inf. Softw. Technol. **53**(10), 1084–1096 (2011)

12. Alsanoosy, T., Spichkova, M., Harland, J.: Cultural influence on requirements engineering activities: a systematic literature review and analysis. Requirements Eng. **25**(3), 339–362 (2019). https://doi.org/10.1007/s00766-019-00326-9

13. Sharma, T., Spinellis, D.: A survey on software smells. J. Syst. Softw. **138**, 158–173 (2018)

14. Beer, A., Junker, M., Femmer, H., Felderer, M.: Initial investigations on the influence of requirement smells on test-case design. In: 25th IEEE International Requirements Engineering Conference Workshops (REW), pp. 323–326. IEEE, Portugal (2017)

15. Bjarnason, E., Unterkalmsteiner, M., Borg, M., Engström, E.: A multi-case study of agile requirements engineering and the use of test cases as requirements. Inf. Softw. Technol. **77**, 61–79 (2016)

16. Mund, J.M.: Measurement-based quality assessment of requirements specifications for software-intensive systems. Doctoral dissertation, Technische Universität München (2017).

17. Toulmin, S.E.: The Uses of Argument. Cambridge University Press, UK (2003)

Author Index

Printed in the United States
by Baker & Taylor Publisher Services